Introduction to Semiconductor Devices

Introduction to Semiconductor Devices

Diodes, Bipolar Transistors, JFETs, IGFETs, SCRs and integrated circuits

F. J. BAILEY

C.Eng., M.I.E.R.E.,
Lecturer in Electronics, Cambridgeshire College of Arts and Technology

London. George Allen & Unwin Ltd
Ruskin House Museum Street

First published in 1972

ISBN 0 04 621017 2 *paper*
0 04 621016 4 *hardback*

Printed in Great Britain
in 10 on 12 pt 'Monophoto' 327 Times
by Page Bros. (Norwich) Ltd., Norwich

For Suzy, Sara and James

Foreword

The invention of the transistor changed the face of the electronic engineering industry within a decade. The subsequent development of field-effect transistors and of integrated circuits has brought about considerable cost and size reduction in electronic equipment but the high rate at which new devices and circuits appear on the market can be very daunting for the engineering technician. How does he decide which device or circuit to use and how does he compare the relative advantages of one device with others which initially appear to be similar? Also, the development of semiconductor materials and devices has been surrounded by a considerable volume of sophisticated solid-state physics and the attendant mathematical theory. The author has written this book for those engineers who wish to read about the characteristics and applications of semiconductor devices but do not wish to approach the subject through this sophisticated theory. Hence the treatment in this book is essentially non-mathematical and directed towards technicians. Classifying devices by their characteristics will enable the reader to make a positive choice of a device that is to perform a particular function. The book closes with details of integrated-circuit techniques that will give the technician an insight into this important new technology.

N. HILLER
Cambridge

Contents

Foreword *page* vii

1 BASIC SEMICONDUCTOR THEORY 15

1. 1 Introduction. 1.2 The silicon atom. 1.3 Structure of a silicon crystal. 1.4 Free electrons. 1.5 Holes. 1.6 Current flow in silicon. 1.7 Impurities. 1.8 n-type silicon. 1.9 Current flow in n-type silicon. 1.10 p-type silicon. 1.11 Current flow in p-type silicon.

2 SEMICONDUCTOR DIODES 26

2.1 Introduction. 2.2 The pn junction. 2.2.1 Potential barrier and depletion layer. 2.2.2 Forward bias. 2.2.3 Reverse bias. 2.3 Symbol for a semiconductor diode. 2.4 Current–voltage characteristic of a semiconductor diode. 2.4.1 Forward characteristic. 2.4.2 Reverse characteristic. 2.4.3 Avalanche breakdown. 2.4.4 Diode resistance. 2.5 Diode or depletion capacitance. 2.6 Equivalent circuit of a diode. 2.7 Zener diodes. 2.7.1 Zener breakdown. 2.8 Applications of junction diodes. 2.8.1 Rectification. 2.8.2 Half-wave rectification. 2.8.3 Full-wave rectification using centre-tapped transformer. 2.8.4 Full-wave rectification using diode bridge circuit. 2.9 Smoothing. 2.10 Application of a zener diode as a voltage regulator. 2.10.1 Zener diode characteristic. 2.10.2 Simple voltage regulation circuit.

3 BIPOLAR TRANSISTORS 50

3.1 Introduction. 3.2 Operation of a transistor. 3.2.1 Transistor action. 3.2.2 Current flow in a npn transistor. 3.2.3 Amplification. 3.2.4 Operation of a pnp transistor. 3.2.5 Current flow in a pnp transistor. 3.3 Transistors in practice. 3.3.1 Common-base (CB) connection and characteristics. 3.3.2 Input characteristic for the CB connection. 3.3.3 Output characteristic for the CB connection. 3.3.4 Transfer characteristic for CB connection. 3.4 Simple CB amplifier. 3.4.1 Voltage and power amplification. 3.4.2 Disadvantages and advantages of the CB amplifier. 3.5 Common-emitter (CE) connection and characteristics. 3.5.1 Input characteristic for the CE connection. 3.5.2 Output characteristic for the CE connection. 3.5.3 Transfer characteristic for the CE connection. 3.6 Simple CE amplifier. 3.7 Advantages and disadvantages of the CE connection.

3.8 Common-collector (CC) connection. 3.9 Equivalent circuit of a transistor. 3.10 Electrical characteristics and maximum ratings. 3.11 Application of a transistor as an amplifier. 3.11.1 The d.c. load line and operating point. 3.11.2 d.c. bias stabilization. 3.11.3 Current, voltage and power gain. 3.11.4 Practical audio-frequency amplifier.

4 JUNCTION FIELD EFFECT TRANSISTORS 93

4.1 Introduction. 4.2 Basic structure of the junction FET. 4.3 Operation of a junction FET. 4.3.1 Absence of bias voltages. 4.3.2 Gate potential but no V_{DS}. 4.3.3 Drain potential but no V_{GS}. 4.3.4 Effect of both V_{GS} and V_{DS}. 4.4 Output characteristic. 4.5 Mutual or transfer characteristic. 4.6 Parameters g_m, r_d and μ. 4.7 Amplification. 4.8 Equivalent circuit. 4.9 Temperature effects. 4.91 Leakage current. 4.9.2 Variation of drain current with temperature. 4.10 Electrical parameters and maximum ratings. 4.11 Junction FETs in practice. 4.11.1 CS amplifier. 4.11.2 CS biasing. 4.11.3 Voltage gain of an audio-frequency CS amplifier. 4.11.4 CD amplifier or source-follower. 4.11.5 Source-follower bootstrap circuit. 4.12 Applications of the junction FET. 4.12.1 Compound or hybrid source-follower. 4.12.2 A junction FET as a voltage variable resistor.

5 INSULATED GATE FIELD EFFECT TRANSISTORS 139

5.1 Introduction. 5.2 Enhancement IGFET. 5.2.1 Operation of the enhancement IGFET. 5.3 Depletion IGFET. 5.3 Depletion IGFET. 5.3.1 Operation of the depletion IGFET. 5.4 Effects of V_{DS} on the channel conductivity of an IGFET. 5.5 Symbols and bias polarities. 5.6 Output characteristics. 5.7 Transfer characteristics. 5.8 Equivalent circuit. 5.9 Electrical parameters and ratings. 5.10 The IGFET as an amplifier. 5.11 Application of an IGFET in an electrometer. 5.12 Application of an IGFET as an electronic switch.

6 SILICON CONTROLLED RECTIFIER (THYRISTOR) 168

6.1 Introduction. 6.2 Operation. 6.2.1 Gate triggering. 6.3 Typical V/I characteristic for the SCR. 6.4 Parameters controlling α. 6.5 Switching on the SCR. 6.6 Gate characteristic. 6.7 Switching off the SCR. 6.8 SCR ratings. 6.8.1 Typical manufacturer's ratings and parameters. 6.9 Applications of the SCR. 6.9.1 The SCR as a static current switch. 6.9.2 Inversion. 6.9.3 Power control. 6.10 Triacs.

7 INTEGRATED CIRCUITS 198

7.1 Introduction. 7.2 Integrated circuit technology. 7.2.1 Purification and preparation of the silicon wafers. 7.2.2 Epitaxial growth. 7.2.3 Oxidation. 7.2.4 Photomasking and etching.

7.2.5 Diffusion. 7.2.6 Metal deposition (interconnections). 7.3 Element isolation. 7.3.1 Diode isolation. 7.3.2 Oxide isolation. 7.4 Formation of integrated circuit elements. 7.4.1 Transistors for integrated circuits. 7.4.2 Diodes for integrated circuits. 7.4.3 Capacitors for integrated circuits. 7.4.4 Resistors for integrated circuits. 7.5 Formation of a simple integrated circuit. 7.6 Field-effect transistors for integrated circuits. 7.7 Encapsulation.

Glossary of Important Terms 232

Index 236

Chapter 1

Basic Semiconductor Theory

1.1 Introduction

A semiconductor is, as its name suggests, a material whose electrical conductivity lies between that of an insulator and a conductor. Germanium and silicon are the two most widely used semiconductors. However, silicon has superior thermal properties compared to germanium, e.g. maximum operating temperature for germanium transistors is 80°C, whereas silicon transistors may be used at temperatures up to 160°C. Silicon is by far the most important semiconductor material used in the present-day manufacture of transistors and other semiconductor devices.

The behaviour of semiconductor devices can best be explained by the aid of simple atomic theory, and since silicon is the basic material used, the following simplified atomic theory is based on silicon.

1.2 The Silicon Atom

The silicon atom, like all other atoms, consists of a central **positively** charged **nucleus**; orbiting around this nucleus in a number of different shells or orbital paths are **negatively** charged particles called **electrons** (Fig. 1.1a). Inside the small dense nucleus are a number of positively charged particles called protons, and the number of these protons inside the nucleus is equal to the number of electrons orbiting around it, i.e. the protons and electrons have equal but opposite charges, resulting in the complete atom being electrically neutral.

The orbiting electrons fill the various shells in definite numbers. The first shell, the one nearest the nucleus, is said to be complete when it is occupied by two electrons, and the second shell is complete when occupied by eight electrons. The third shell is conditionally filled by eight electrons and completely filled by eighteen, the fourth shell is conditionally filled by eight or eighteen electrons and completely filled by thirty-two, and so on. The electrons in a normal unexcited atom arrange themselves such that the only shell which may be incomplete is the outer one, and the properties of an element are very largely determined by the conditions existing in this shell. Electrons which occupy the outer shell are called **valence electrons.**

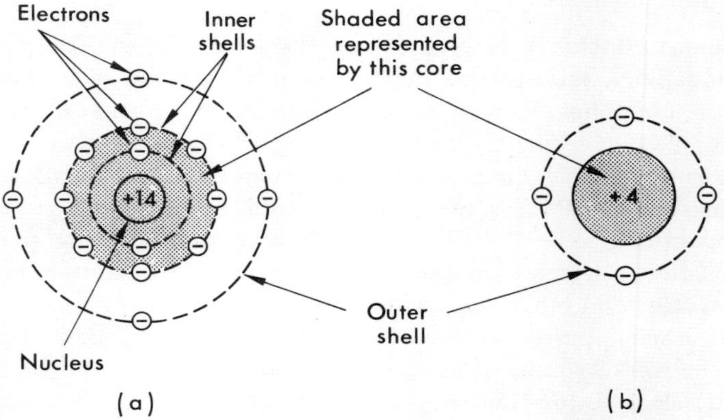

Fig. 1.1 Representation of a silicon atom.
(a) Silicon atom.
(b) Simplified diagram of a silicon atom.

If the outer shell of an atom is complete as in inert gases, then the atom is very stable and will not combine chemically with other atoms. If the outer shell of an atom is incomplete, as is the case with semiconductors, then the atoms will combine chemically with each other in such a way that by sharing valence electrons the outer shells will appear to have their full complement of electrons and so be conditionally stable.

Each negatively charged electron is held in its respective shell by the balance which exists between the centrifugal force on the orbiting electron and the attractive inward force of the positive nucleus. The closer an electron is to its nucleus, the more tightly is it held in its particular orbit. Thus valence electrons in the outer shell are held less firmly than those in the inner shells.

The electrons in the outer shell of a conductor are very loosely held in orbit, so much so that at normal room temperature some of these outer electrons leave their parent atom and wander freely from atom to atom. Such electrons are called **free electrons,** and under the influence of a small applied potential difference, these free electrons are attracted towards the positive potential—so constituting an electric current.

In an insulator, electrons in the outer shell are much more tightly held in their shells by the nucleus so that there are no free electrons. However, if a large enough potential difference is applied across an insulator, some valence electrons will be forced out of their shells by the attractive force of the applied electric field and will flow towards the positive potential, so forming a small electric current.

The silicon atom has three shells of orbiting electrons. The inner two have their full complement of electrons and are relatively unimportant, but the outer shell, which has only four valence electrons instead of the eight required for completeness, is vitally important. Since it is this outer shell with its four valence electrons that is so important, the silicon atom may be represented more simply by Fig. 1.1(b). The central core represents the nucleus with its fourteen positively charged protons, and the two inner shells contain ten negatively-charged electrons. The net charge of the core will be $+14-10$ which gives $+4$. Since the net charge of an atom is zero, the $+4$ of the core is balanced by the four negatively-charged valence electrons shown orbiting the core.

1.3 Structure of a Silicon Crystal

Fig. 1.2 illustrates the structure of a perfect silicon crystal. In reality the structure is three-dimensional. The atoms arrange themselves into a regular pattern called a **crystal lattice** where the bonds or ties between the atoms are formed by the four valence electrons of one atom interchanging with the four valence electrons from its four neighbouring atoms. It must be understood that the atoms are locked into place within the crystal by these four valence electrons interchanging between their parent atom and those of neighbouring atoms. The paths of interchange are called **covalent bonds.**

1.4 Free Electrons

If all the valence electrons were engaged in covalent bonds, silicon would be a perfect insulator, i.e. there would be no free electrons available to constitute an electric current, but this would only occur at absolute zero temperature (−273°C). At room temperature some valence electrons absorb enough heat energy to enable them to break free from their covalent bonds. When this occurs, the liberated electrons are free to wander throughout the crystal. If a battery is connected across the silicon crystal, these free electrons

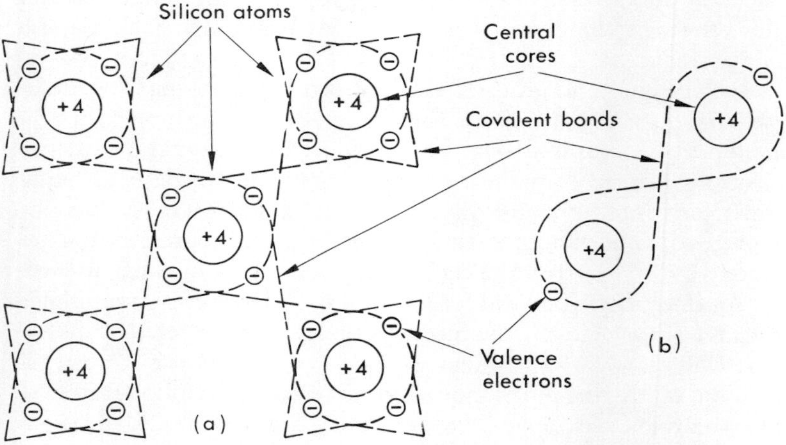

Fig. 1.2 Bonding of silicon atoms by valence electrons.
(a) Structure of silicon crystal.
(b) Two silicon atoms shown sharing two valence electrons.

are attracted to the positive terminal and so constitute an electrical current. The higher the temperature, the greater will be the number of free electrons and consequently the greater the resulting current.

1.5 Holes

Whenever an electron succeeds in breaking its covalent bond within a semiconductor crystal and wanders off, it leaves behind an electron vacancy. This vacancy is deficient of a negative charge, and therefore it must appear to the surrounding electrons as an equal and opposite positive charge. Such regions are conveniently thought of as **positively-charged holes.** This concept of holes, which at first may seem odd, has been universally adopted and the reader must be prepared to accept it.

By virtue of their positive charge (unlike charges attract one another), holes are capable of attracting and causing some neighbouring electrons to break their covalent bonds. Thus an electron may leave a covalent bond at point A in the crystal and fill a hole point B. The electron, on leaving point A, leaves behind a hole and on reaching point B it occupies one. Consequently it appears as if the hole itself has moved from B to A. It is by this process that holes appear to move in a stop-start fashion as a negative charge deficiency or as an effective positive charge.

At room temperature, a large number of covalent bonds is always being broken, and each time a bond is broken an **electron–hole pair** is produced. Should a voltage be applied across the silicon crystal, both the electron and the hole will contribute to the resultant current flow.

Although electrons and holes can co-exist within a crystal, holes are being continually filled by electrons and this process is known as **recombination.** For a given temperature the number of electrons and holes within the crystal is. practically constant, because on average there are as many electrons and holes lost due to recombination as there are produced by electron–hole pair generation.

1.6 Current Flow in Silicon

If a voltage is applied across a small block of pure silicon, a current will flow (Fig. 1.3). This current will consist of two different types of carrier:

(a) Free electrons—these are negatively-charged carriers.
(b) Holes—these are positively-charged carriers.

The positive terminal of the battery will attract the electrons (unlike charges attract one another) and repel the holes (like charges repel one another). Conversely the negative terminal will attract the holes and repel the electrons. Thus the free electrons drift towards end A, while the holes drift in a stop–start fashion towards B.

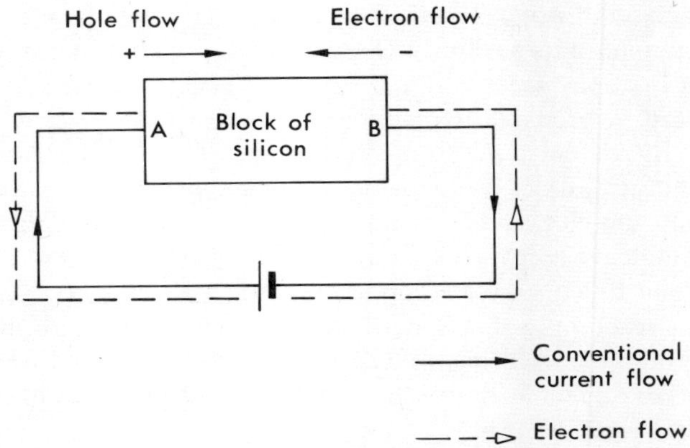

Fig. 1.3 Electron and hole flow in silicon.

Although under the influence of the battery a current will flow around the closed circuit, the number of electrons and holes within the silicon block will always remain constant. An electron, on arriving at end A, leaves the block of silicon and flow into the positive terminal of the battery. But for every electron such as this that enters the positive terminal of the battery, another electron will leave the negative terminal and flow into end B of the silicon, i.e. the number of electrons in the silicon is constant. Furthermore for every hole that arrives at end B of the silicon, an electron leaves the negative terminal of the battery and recombines with it. For every hole which is lost in the silicon in this way, a new hole is created at end A by an electron leaving the silicon and flowing into the positive terminal of the battery.

Hence it may be seen that the total current flowing is the sum of the electron and hole flow. Further, that although within the silicon block there are two different types of current carrier, i.e. electrons and holes, there is only the electron flow in the external connecting wires, since holes cannot exist outside a semiconductor material.

1.7 Impurities

Although the number of current carriers increases with temperature, a more exact method of producing more free electrons or holes is by the addition of a minute quantity of an impurity to the pure silicon. This is known as **doping,** and the **doping level** refers to the ratio of impurity atoms to silicon atoms. For a transistor this may typically be $1:10^8$, i.e. one impurity atom for every hundred million silicon atoms. Since there are billions upon billions of silicon atoms in the volume of a pinhead, the above ratio indicates a very large number of impurity atoms in one transistor device.

The added impurity atoms, which are distributed throughout the silicon, may be divided into two groups:

(a) **Pentavalent impurity;** this type of impurity has **five** valence electrons in the outer shell of its atom. Important examples are arsenic and antimony.
(b) **Trivalent impurity;** This type has **three** valence electrons in the outer shell of its atom. Important examples are gallium and indium.

1.8 n-type Silicon

Fig. 1.4 shows a pentavalent impurity atom in a crystal of silicon. For simplicity only five atoms have been shown; four silicon atoms surrounding a single pentavalent impurity atom. The diagram illustrates that four of its five valence electrons have entered into covalent bonds with four neighbouring silicon atoms, but the fifth electron is in excess to requirements for covalent bonding. Consequently it is free and is available as a current carrier. Since the pentavalent impurity atom has donated a free electron it is often called a **donor impurity.** Thus, by the addition of donor impurities to the silicon, large numbers of free electrons become available as current carriers. Since conduction will be predominantly by electrons, and since electrons are negatively charged, the material so created is known as **n-type silicon.**

Note that the inner part of the impurity atom, called the core, is represented by a circle containing $+5$. This is so because the pentavalent impurity atom can be represented by a core, with five valence electrons orbiting around it. Thus the core (+5) and five valence electrons (−5) make the net charge of the atom zero, i.e. neutral. However, the impurity atom donated a free electron and in so doing it must become positively charged, since it was neutral before losing a negatively charged electron. When an atom loses an electron it is called a **positive ion.**

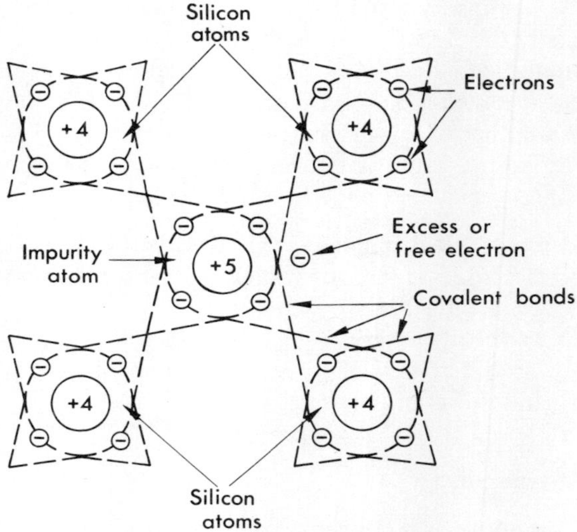

Fig. 1.4 n-type silicon.

Thus pentavalent or donor impurity atoms within the crystal become positive ions. Although the donated electrons are free to wander within the crystal, the positive ions always remain fixed or immobile within the crystal lattice held by their covalent bonds.

1.9 Current Flow in n-type Silicon

If a voltage is applied across a block of n-type silicon, then the resulting current will consist essentially of free electrons. However, at room temperature there would also be a small number of holes produced by electron–hole pair generation. Under the influence of the battery voltage the electron and hole flow would be as indicated in Fig. 1.5. The electrons, being present in far greater numbers than holes, are called the **majority carrier,** while the holes are called the **minority carrier.**

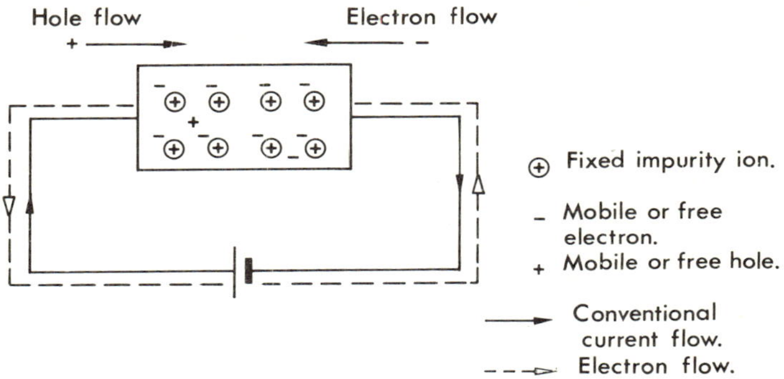

Fig. 1.5 Current flow in n-type silicon.

1.10 p-type Silicon

Instead of adding a pentavalent impurity, a trivalent impurity may be added to the pure silicon. The three outer electrons of the impurity atom enter readily into cavalent bonds with three neighbouring silicon atoms, but in order to form a four-electron covalent bond it has to 'rob' a neighbouring atom to obtain the extra electron. Hence, the impurity atom 'accepts' an electron from a silicon atom, so causing the silicon atom to have an electron vacancy or a hole. Due to this robbing or accepting process, trivalent impurities are known as **acceptor impurities.**

Thus by the addition of acceptor impurities, holes are created and current flow will be predominantly by holes. Since holes are positive, the material so formed is called **p-type silicon.**

Note that in the case of trivalent or acceptor impurities, the core is represented by a circle containing $+3$. This must be so because there are only three electrons in the outer shell, and the core $(+3)$ with the three valence electrons (-3) give a net charge of zero. Now when the trivalent impurity atoms are added to silicon, they always succeed in accepting an electron from neighbouring silicon

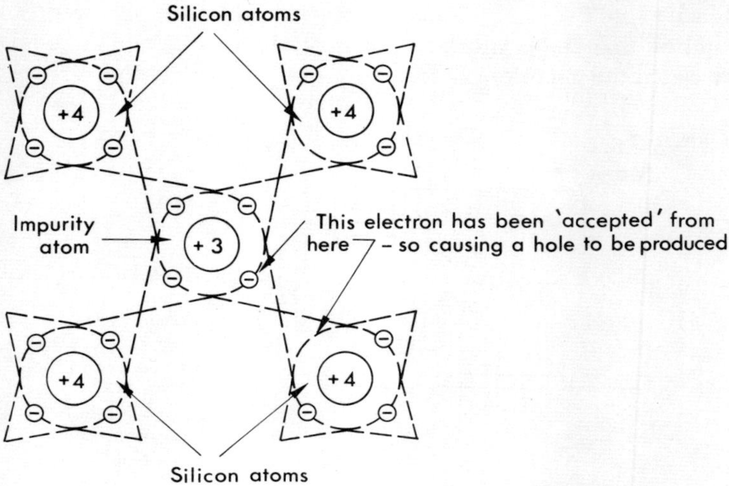

Fig. 1.6 p-type silicon.

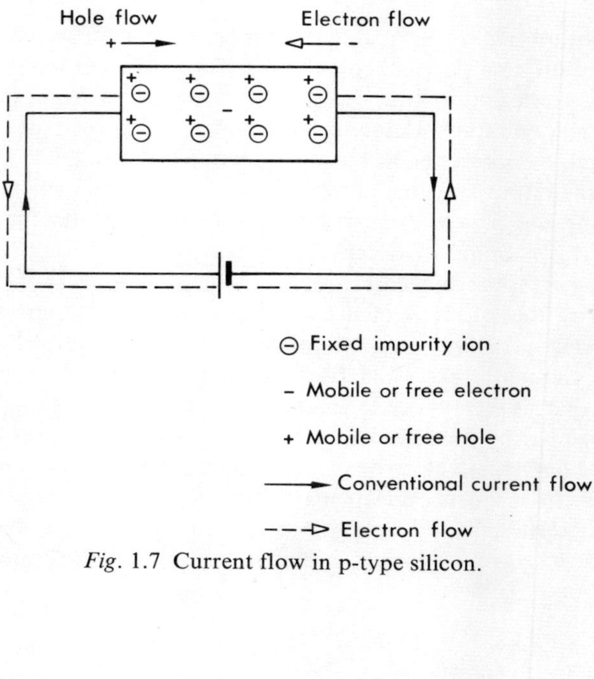

Fig. 1.7 Current flow in p-type silicon.

atoms. Thus they acquire an extra electron above their normal number. Atoms which gain an extra electron are called **negative ions.**

So trivalent or acceptor impurities become negative ions, and like their counterparts, the positive ions in n-type silicon, they remain fixed in the crystal whereas the holes they produce are mobile and free to wander.

1.11 Current Flow in p-type Silicon

Under the action of an applied voltage, the holes, which are the majority carriers, and the electrons, which are the minority carriers, flow in the directions indicated in Fig. 1.7.

REFERENCES

ADER, R. B. and R. L. LONGINI. *Introduction to Semiconductor Physics*, John Wiley and Sons Inc., 1963.

PRIDHAM, G. J. *Electronic Devices and Circuits*, Pergamon Press Ltd., 1968.

SHIRE, J. N. *Physics of Solid State Electronics*, Charles E. Merrill Books Inc., 1966.

SHOCKLEY, W. *Electrons and Holes in Semiconductors*, D. Van Nostrand Company Inc., 1950.

Chapter 2

Semiconductor Diodes

2.1 Introduction

The operation of practically all solid state devices, e.g. pn diodes, transistors, silicon controlled rectifiers, field-effect transistors, etc., depend upon the action of a pn junction. Consequently, the theory outlined in this chapter is not only relevant to the diode, but also to all subsequent semiconductor devices described in this book. Therefore a full understanding of the contents of this chapter will help considerably in mastering the behaviour of other more complicated devices.

2.2 The pn Junction

2.2.1 *Potential barrier and depletion layer*

A semiconductor diode is formed whenever p-type and n-type material are brought together to form a pn junction. When such a junction is made, electrons and holes which happen to be close to the junction are mutually attracted and cross over and recombine. This recombination leaves a very thin region depleted of electrons and holes on either side of the junction. The loss of the holes in the p-region exposes negatively charged acceptor ions, whilest in the n-region the loss of electrons exposes positively charged donor ions (Fig. 2.1). Owing to the presence of these exposed charges the migration of carriers across the junction is very quickly stopped. These charges act as though a very small internal battery were connected across the junction. The polarity of this battery will stop the flow of majority carriers across the junction, because the negative potential in the p-region repels the electrons in the n-region, while the positive potential in the n-region repels the holes from crossing into the n-region.

Because this apparent internal battery blocks majority carriers from crossing the junction it is called an *internal potential barrier.*

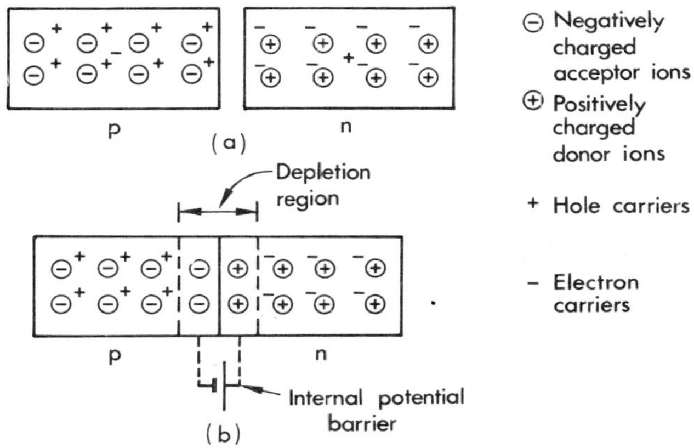

Fig. 2.1 Depletion region and potential barrier of a pn junction.
(a) Separate p- and n-type silicon.
(b) pn junction.

Its potential difference is only a few tenths of a volt. Note that although the potential barrier blocks majority carriers from crossing the junction, its polarity is such as to attract minority carriers to cross. Since there are no carriers within the region where the internal potential barrier is developed except minority carriers in transit, it is called the **depletion layer,** i.e. depleted of carriers. This depletion layer is very thin, being less than one micron (μm) wide.

2.2.2 *Forward bias*

When an external voltage is applied across the junction, as shown in Fig. 2.2, the applied voltage acts in opposition to the internal potential barrier. Two important effects take place. Initially the applied positive potential will repel the holes in the p-material towards the junction, whilst the applied negative potential will repel the electrons in the n-material also towards the junction. The result is that the small internal potential barrier is neutralized or overcome. Consequently the majority carriers can now cross over the junction and their flow constitutes a current.

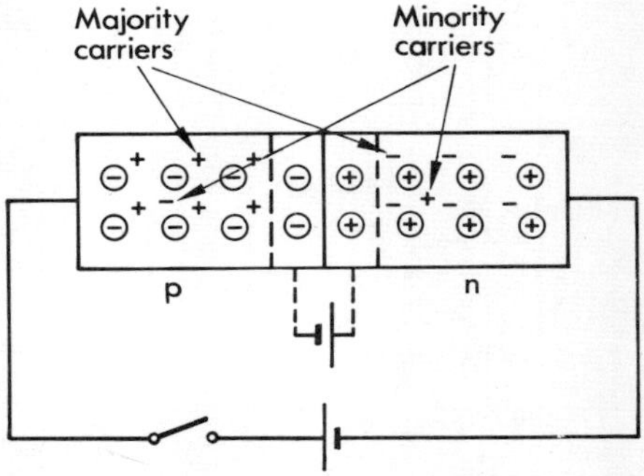

Fig. 2.2 pn diode with forward bias.

The internal potential barrier is neutralized because the donor and acceptor ions forming the barrier region loose their charges. For example, if a positive donor ion within the depletion layer in the n-region of the junction regains an electron, its net positive effect to the potential barrier is lost. The same may be said for a negative acceptor ion. If it regains a hole, then its negative effect to the potential barrier is lost. Thus, if a large number of ions within the depletion layer recapture their lost charges due to carriers being forced towards the junction by the external voltage, the result is a lowering of the internal potential barrier. Furthermore, because the ions on the outside of the depletion layer regain their charges first, the initial effect is that the width of the depletion layer is reduced (Fig. 2.3). If the external voltage is large enough, the depletion layer shrinks further and disappears, removing the internal potential barrier. As a result of this external voltage, the majority carriers can flow across the junction and current is said to flow in the forward direction. The pn junction is said to be biased in the **forward direction.**

Although majority carriers flow across the junction, the number of holes and electrons within the p- and n-regions always remains constant. This is so, because, for every hole which leaves the p-region

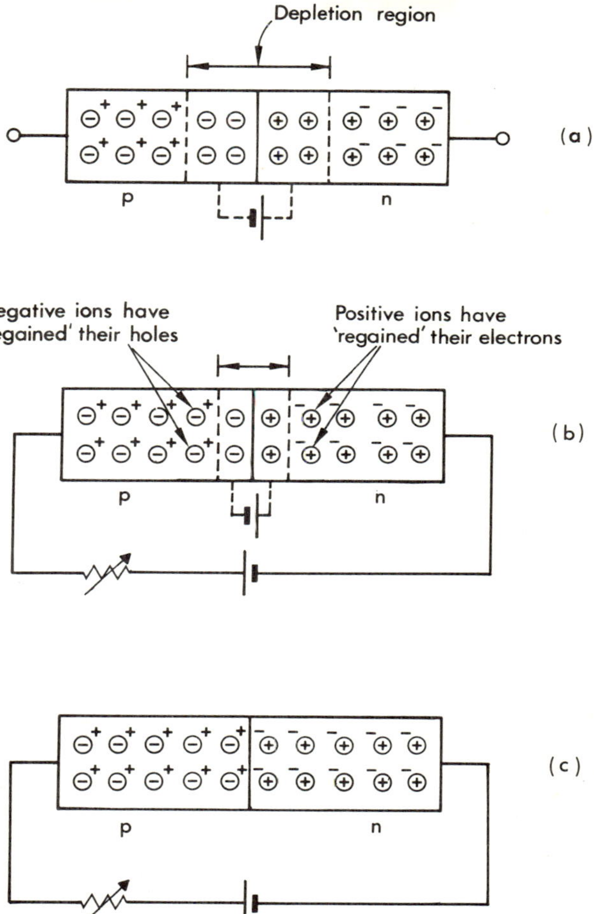

Fig. 2.3 The effect of forward bias on the depletion region.

(a) No bias.

(b) Small forward bias, depletion region width reduced but not eliminated.

(c) Forward bias increased so that the depletion region and its associated internal potential barrier have been neturalized.

and recombines with an electron in the n-region of the junction, another hole is created within the p-region by an electron leaving it and flowing to the positive terminal of the external battery. Similarly, any electrons lost in the n-region are replaced by electrons

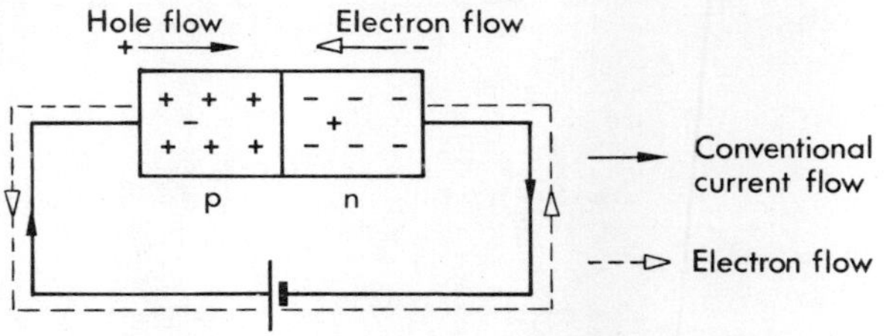

Fig. 2.4 Current flow in a forward-biased pn diode.

leaving the negative terminal of the battery and flowing into the n-region. Thus the battery compensates for electrons and holes which are lost due to recombination. Hence in the external circuit the total current flowing is made of electrons only, whereas in the diode the current flow is by electrons in the n-region and holes in the p-region. This is illustrated in Fig. 2.4.

2.2.3 *Reverse bias*

If the polarity of the external battery is reversed, it will be connected

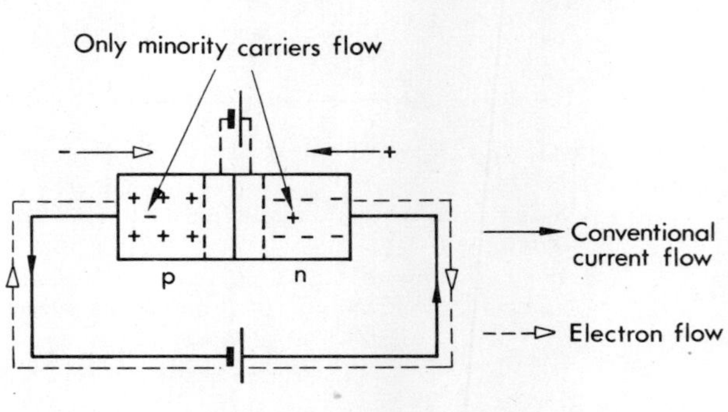

Fig. 2.5 pn diode with reverse bias.

so as to aid the internal potential barrier, and consequently increase its effect. The majority carriers will be unable to flow across the junction and the junction is said to be **reverse biased** (Fig. 2.5).

Although the majority carriers are blocked by the reverse biasing and the internal potential barrier, their polarities are such as to attract the minority carriers across the junction. Therefore under the action of reverse biasing, a small current will flow through the diode made up entirely of minority carriers which are being continually produced by electron–hole pair generation (electrons absorbing heat energy and breaking their covalent bonds).

On reverse-biasing the pn junction, majority carriers are attracted away from the junction. As the carriers are swept away, positive and negative ions are exposed due to losing their associated carriers (Fig. 2.6). The result is that the depletion layer width is increased by the addition of these ions, which also means that the internal

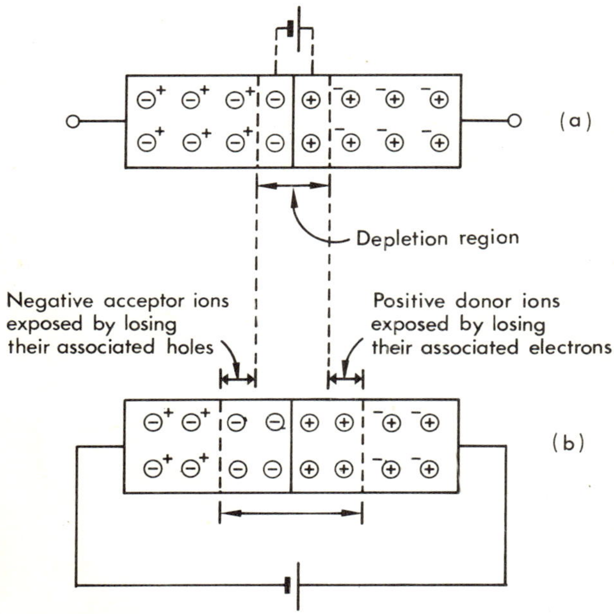

Fig. 2.6 Widening of depletion layer by reverse bias.
(a) No bias voltage.
(b) Reverse bias voltage applied, depletion region width increased.

potential barrier is increased. Thus the width of the depletion layer is a function of the magnitude and polarity of the voltage across the junction.

2.3 Symbol for a Semiconductor diode

The symbol for a semiconductor diode is shown in Fig. 2.7(b). The

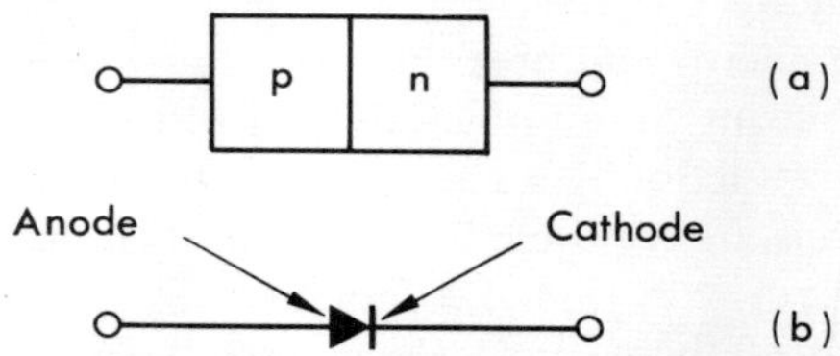

Fig. 2.7 Symbol of a semiconductor junction diode.
(a) Structure.
(b) Graphical symbol.

anode is always the p-type material, while the cathode is the n-type, the triangle points in the direction of conventional current flow.

2.4 Current-Voltage Chasacteristic of a Semiconductor Diode

Fig. 2.8 illustrates a characteristic relating current and voltage for low-power semiconductor diode.

2.4.1 *Forward characteristic*

When the diode is forward biased and the voltage increases from zero, only a relatively small current flows (region OA). This is because the external voltage is being opposed by the internal potential barrier. However, as soon as the internal potential barrier is neutralized, current through the diode increases rapidly (region AB).

As the forward bias voltage is increased, the electron carriers travel faster and their kinetic energy increases so that when they collide with other atoms they are capable of dislodging electrons from their covalent bonds; this creates further electron–hole pairs which add to the current flow. Furthermore, as the current increases the heating effect produced by it increases; this heat is dissipated in the semiconductor material, so causing its temperature to increase. Since the generation of electron–hole pairs is temperature dependent, further current carriers are produced which add to the total current flowing. This whole effect is cumulative, so care must be taken that the maximum current rating for the diode is not exceeded, otherwise permanent damage to the diode will result.

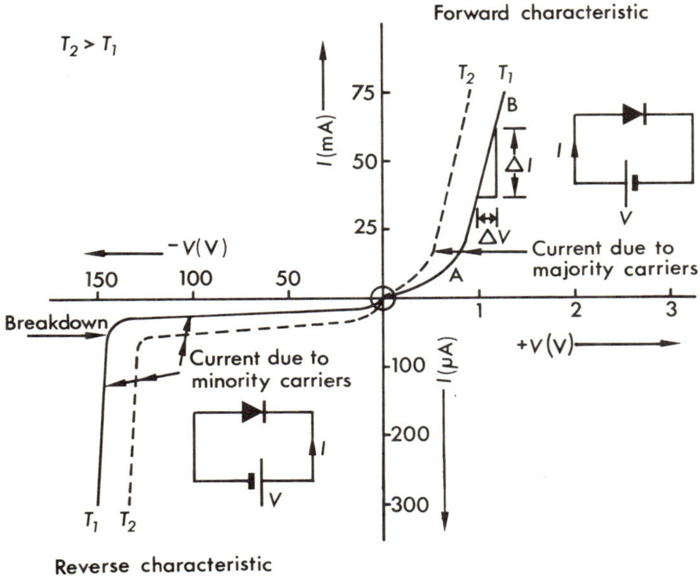

Fig. 2.8 *I/V* characteristic of a semiconductor diode.

The current through a semiconductor diode is temperature sensitive and this effect is illustrated in Fig. 2.8 by the *I/V* characteristic for two temperatures, where T_2 is greater than T_1.

2.4.2 *Reverse characteristic*

When the diode is reverse biased, majority carriers are blocked and only a small current (minority carriers) flows through the diode. As the reverse voltage is increased from zero, the reverse current very quickly reaches its maximum value, or saturation level. This occurs when the rate of flow of minority carriers across the junction is equal to the rate at which they are produced by thermal breakdown. If the temperature is increased from T_1 to T_2 the rate of electron–hole pair generation increases, and so the reverse saturation current increases.

The reverse saturation current is often referred to as the diode's **leakage current.**

From Fig. 2.8 it may be seen that if the reverse bias voltage is increased to a critical value, the reverse current increases rapidly and the junction is said to break down. There are two different types of breakdown, avalanche and Zener. The latter will be discussed in the section on Zener diodes.

2.4.3 *Avalanche breakdown*

As the reverse voltage is increased, the internal potential barrier is increased. As a result the minority carriers are attracted across the depletion layer so quickly that at a critical value of reverse bias they attain sufficient kinetic energy to produce further electron–hole pairs during their frequent collisions with atoms. The newly generated current-carriers may also be accelerated so that they, too, produce on collision, further electron–hole pairs. The action is cumulative (like an avalanche) resulting in a large reverse current. If the current is not limited by some external circuit resistance the diode may be permanently damaged.

2.4.4 *Diode resistance*

At any point on the I/V characteristic (Fig. 2.8) the ratio

$$\frac{\Delta V}{\Delta I} = \frac{\text{Small change of voltage}}{\text{Corresponding change in current}}$$

is the a.c. resistance of the diode.

For the forward characteristic the resistance R_f is relatively small (10–500 Ω) but for the reverse characteristic it is large, e.g. R_b can be from 100 kΩ to several megohms. This reverse resistance is usually referred to as the back resistance, hence the symbol R_b.

2.5 Diode or Depletion Capacitance

A capacitor is created when two conducting material are insulated from one another by a dielectric. The capacitor has the property of storing charge as an excess of electrons on one plate and a deficiency on the other.

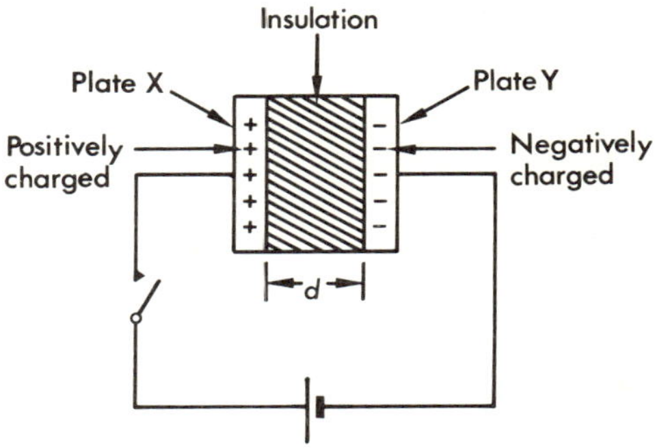

Fig. 2.9 A parallel plate capacitor.

Fig. 2.9 shows a parallel plate capacitor connected across a battery. When the switch is closed a brief charging current flows which causes plate Y to become charged negatively with respect to plate X. The amount of capacitance C between plates X and Y is dependent on the area A of either plate and the distance d between them, i.e.

$$C \propto A/d$$

Consider the semiconductor diode shown in Fig. 2.10. It may be

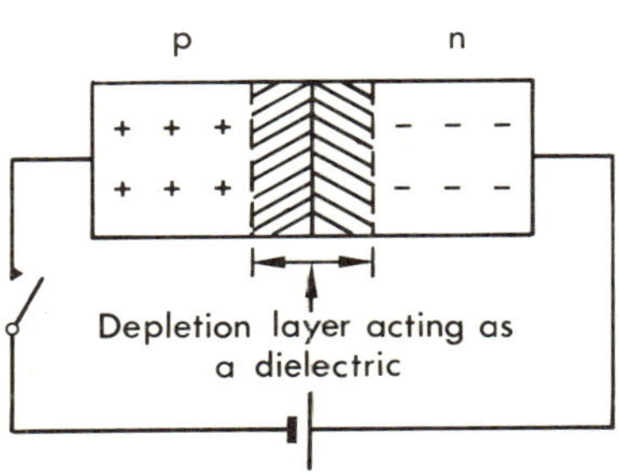

Fig. 2.10 Depletion capacitance.

seen that in effect a parallel plate capacitor is formed in the device. The capacitance is produced by the depletion layer which, being carrier free, acts as an insulator or dielectric between two regions which contain a large number of carriers, i.e. holes in the p-region and electrons in the n-region.

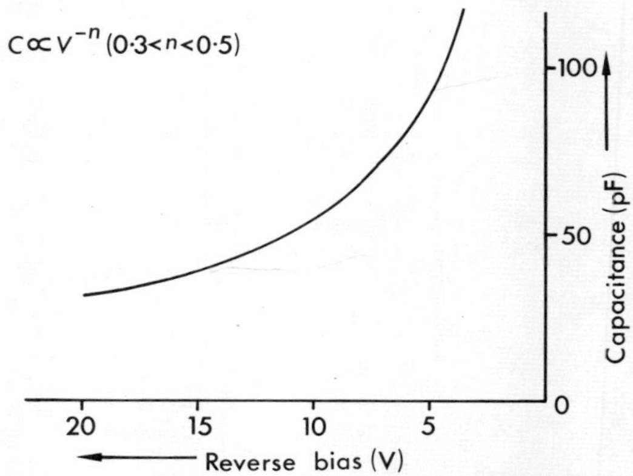

Fig. 2.11 Depletion capacitance as a function of reverse bias.

Increasing the reverse voltage across a pn junction increases the depletion layer width. This will result in an increase in the distance d between the 'plates', and consequently the capacitance decreases. Fig. 2.11 shows the relationship between depletion capacitance and reverse bias for a small junction diode. A junction diode which is used to operate like a variable capacitor is called a **varactor diode.**

2.6 Equivalent Circuit of a Diode

From previous considerations it is apparent that the diode exhibits different electrical characteristics under different operating conditions. For example, when forward biased it appears as a low resistance, whereas when reverse biased it appears as a high resistance shunted by a small depletion capacitance. It is possible and often helpful to represent these different electrical characteristics by an equivalent circuit consisting of resistors and capacitors. The degree of complexity of the equivalent circuit is determined by the particular type of application under consideration.

When the diode is being used for converting a.c. to d.c., it may be considered to be equivalent to a simple electrical switch (Fig. 2.12a). This assumes an ideal diode, i.e. when the diode is forward-biased its resistance R_f is negligible and it passes current like a closed switch. Conversely when the diode is reverse-biased its resistance R_b is assumed infinite and no current is passed, as would be the case for an open switch.

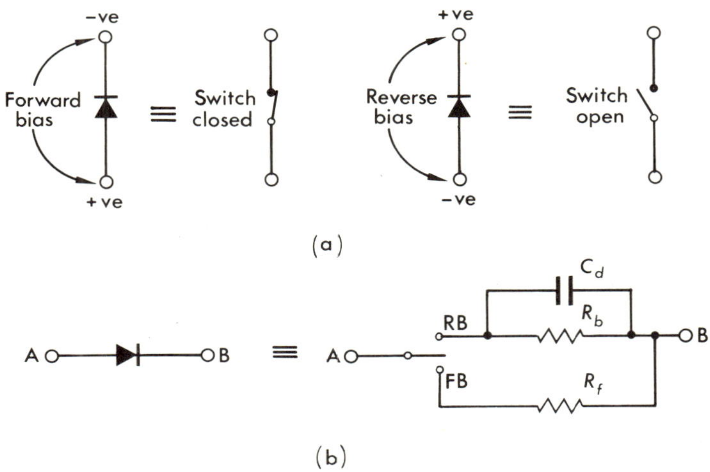

Fig. 2.12 Equivalent circuit of a diode.
(a) The diode as a switch.
(b) A more complete equivalent circuit of a diode.

A more complete equivalent circuit of the diode, which would be useful for small-signal, high frequency work, is shown in Fig. 2.12(b). It may be considered to consist of two parts:

(a) When the diode is forward biased, (FB) it merely appears as a small resistance R_f. (Switch in position FB).
(b) When reverse biased (RB) the diode appears as a high resistance R_b shunted by the depletion capacitance C_d. (Switch position RB).

2.7 Zener Diodes

The junction diode discussed so far may be expected to work alternatively on either the forward or the reverse part of its characteristic. Care is taken to ensure that the maximum forward current or the reverse breakdown voltage is not exceeded, but otherwise the junction diode may be expected to work within these two limits.

Zener diodes, however, are operated exclusively on the reverse characteristic and in particular on the breakdown part of the characteristic. No permanent damage to the device occurs providing its maximum reverse current is not exceeded. The breakdown voltage, is dependent on the diode's construction and it can be made to have individual breakdown voltages in the range 2–200 V, irrespective of current. It is this property which makes the Zener diode so widely used as a voltage regulator or stabilizer.

2.7.1 *Zener breakdown*

Valence electrons may break their covalent bonds in a number of different ways, thereby generating electron–hole pairs, for example, by absorbing heat energy or by being knocked out by other colliding electrons. A third method is by direct disruption of the valence bond by a strong electric field.

Note: Just as there exists a magnetic field between a north and a south pole of a magnet, there also exists an electric field between positive and negative charges. Lines of electric flux making up the electric field go from a positive to a negative charge, and the strength of the field is equal to the potential gradient. For a parallel-plate capacitor where the electric field in the dielectric is uniform, the potential gradient is obtained from the potential difference divided by the distance between the two plates. Considering the capacitor shown in Fig. 2.9, the strength of the electric field between the two plates X and Y is V/d volt/metre.

With Zener diodes the transition from p-type material to n-type is made very abruptly so that the depletion layer may be only 5×10^{-8}m wide. Consequently with a reverse voltage across it of, say, 5 V, the strength of the electric field across the depletion layer will be $5/5\times10^{-8}$ V/m, i.e. 10^8 V/m. This field is sufficiently strong to attract the negatively charged valence electrons directly out of their covalent bonds, so producing a large number of electron–hole pairs which cause the reverse current to increase rapidly.

From the previous explanation of avalanche breakdown, it might be thought that a reverse voltage which gives an electric field of 10^8 V/m may be sufficiently large to give avalanche breakdown as well. But this is not necessarily so, since with Zener diodes the depletion layer width is so very thin that the electrons crossing the layer are not subjected to the accelerating voltage for long enough to attain the required liberating kinetic energy.

There are several distinguishing features between avalanche and Zener breakdown, apart from the required physical details of the junction. The two most important ones are:

(a) Zener breakdown voltage decreases with temperature, whereas avalanche breakdown voltage increases.
(b) Zener breakdown has a sharper turnover characteristic, and once breakdown has occurred, current increases more rapidly for a given increase in voltage.

2.8 Applications of Junction Diodes

Semiconductor diodes are available in a very wide range of current ratings, e.g. maximum forward currents of milliamperes (milli = 10^{-3}) to hundreds of amperes. On account of numerous advantages over their thermionic counterpart, the diode valve, the latter is little used nowadays. Some of the important advantages of junction diodes over valve diodes are:

(a) For a given power rating, their volume and mass is much less.
(b) For small signal diodes, no holder is required to support them in a circuit.
(c) Since they require no heater current, this common source of hum is eliminated.
(d) A considerable saving in power consumption is achieved.

2.8.1 *Rectification*

Most electronic apparatus is d.c. operated. Consequently it is frequently necessary for a.c. mains voltages to be transformed to a suitable value and then converted to d.c. by rectification. Figures

2.13, 2.14 and 2.15 show the three basic rectification circuits which may be employed.

2.8.2 *Half-wave rectification*

This is the cheapest and simplest rectification circuit used. To understand its action, it is best to consider the diode as acting like a switch. When the diode is forward biased it passes current readily, so the switch is closed; conversely when the diode is reverse biased current flow is blocked, so the switch is open (Fig. 2.12).

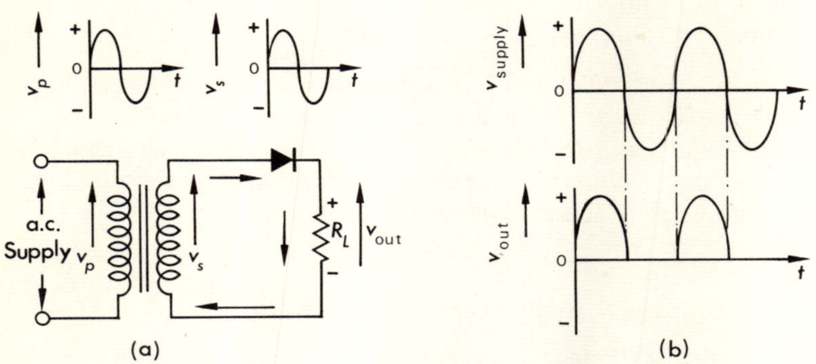

Fig. 2.13 The half-wave rectifier.
(a) Circuit.
(b) Input and output voltage waveforms.

On inspection of Fig. 2.13 it may be seen that as the secondary voltage of the transformer is on the positive half-cycle the diode is forward biased (the anode is positive with respect to the cathode), the switch is closed and current passes through the load and develops a corresponding half-cycle voltage. When the secondary voltage is on the negative half-cycle the diode is reverse biased, the switch is open, and no current reaches the load. Hence the current through the load consists of a series of positive half-cycles, the negative half-cycles of the supply having been blocked or eliminated. Thus the voltage across the load is unidirectional and it is said to be rectified, although as yet it is by no means smooth. In order to obtain smooth direct current, suitable filtering circuits must be employed and they will be explained later.

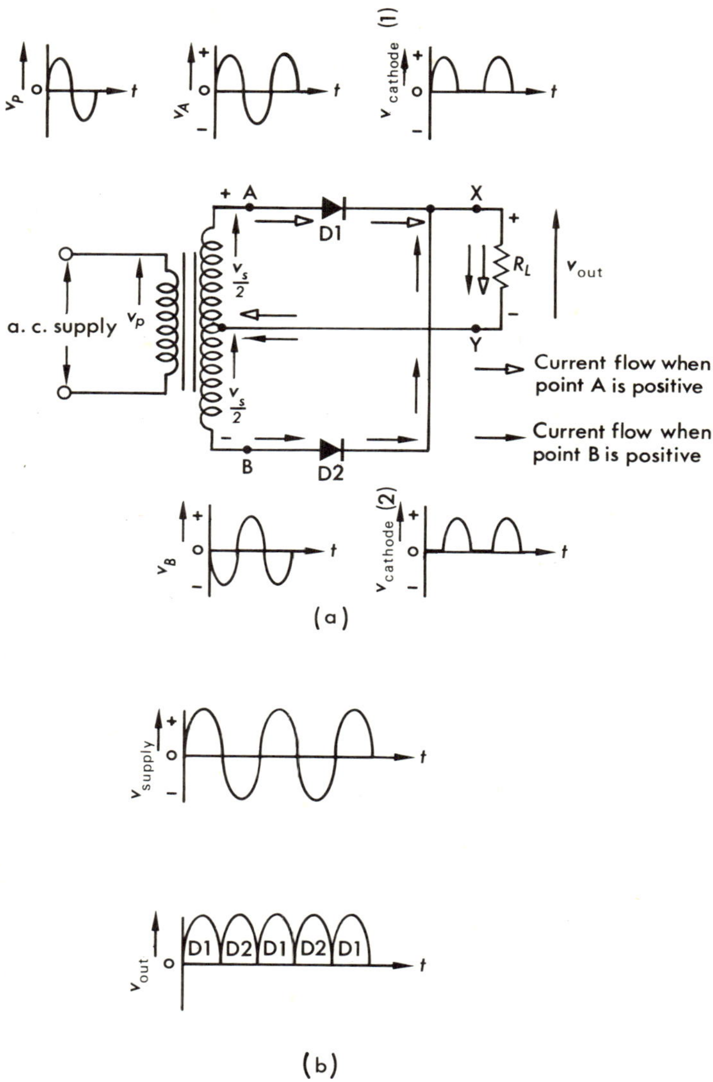

Fig. 2.14 Full-wave rectification using centre-tapped transformer.
(a) Circuit.
(b) Input and output voltage waveforms.

2.8.3 *Full-wave rectification using a centre-tapped transformer*

This circuit, which is shown in Fig. 2.14, is very widely used. Its most serious disadvantage is the expensive centre-tapped transformer necessary to supply voltages, which are equal and opposite, to the two diodes D1 and D2.

The actual voltage waveforms which would be seen by an oscilloscope at certain key points are indicated in Fig. 2.14. It will be noticed that when the anode of D1 is on the positive half-cycle, the anode of D2 is on the negative half-cycle. Thus the voltages at the anodes of the two diodes are 180° out of phase.

Since the diode acts like a closed switch only when its anode is positive with respect to its cathode, it follows that D1 will pass current to the load on the first half-cycle, while D2 acts as an open switch. Conversely on the second half-cycle D2 will pass current to the load while D1 acts as an open switch. Thus D1 and D2 each conduct only an alternative half-cycles and in so doing permit current to flow through the load in one direction only.

The load current is unidirectional as it was with half-wave rectification. But instead of the load receiving only alternate positive half-cycles, it now receives a continuous train of positive half-cycles. The result is that the circuit is operating at twice the efficiency. Furthermore, eventual filtering enables a smooth direct current to be obtained more easily.

2.8.4 *Full-wave rectification using diode bridge circuit*

On comparing the bridge circuit with the previous circuit (Fig. 2.14) it may be seen that the bridge circuit eliminates the need for a centre tapped transformer but incorporates two extra diodes. For normal a.c. mains rectification, the cost of the extra two diodes is less than the difference in cost between an untapped and an accurately centre-tapped transformer. This, and other advantages, have led to the bridge circuit becoming the most widely used for low power rectification.

The circuit operates as follows. When point A is positive with respect to point B, diodes D3 and D4 are reverse biased and so act as open switches. But diodes D1 and D2 are forward biased and so pass current around the circuit as shown by the arrows with the

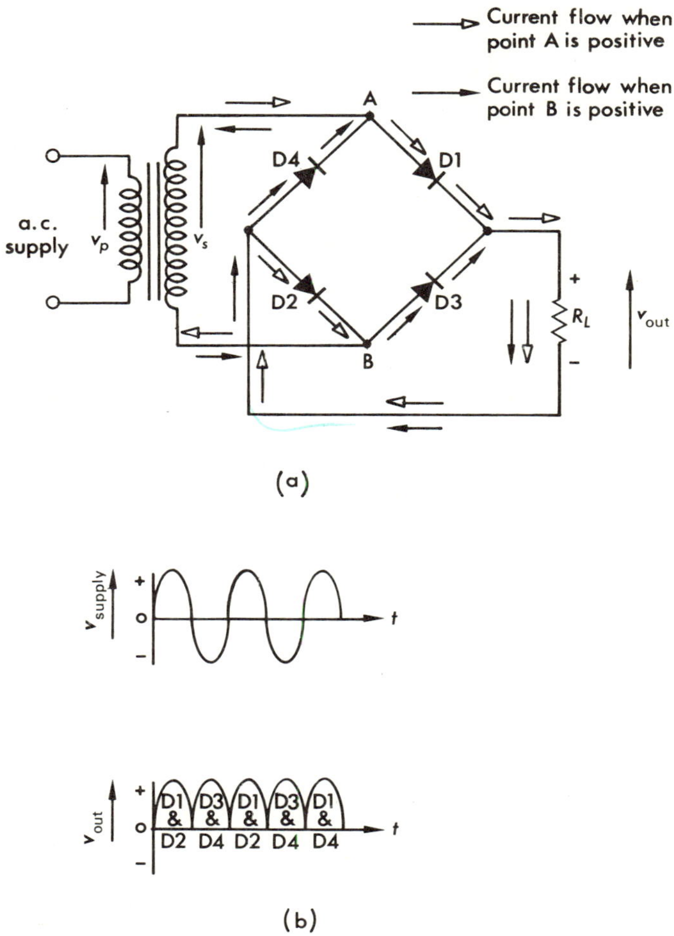

Fig. 2.15 Full-wave rectification using a diode bridge circuit.
(a) Circuit.
(b) Input and output voltage waveforms.

unshaded heads. Conversely, when point B is positive with respect to point A, diodes D1 and D2 become reverse biased and act as open switches. But diodes D3 and D4 are now forward biased and so pass current around the circuit as shown by the arrows with the shaded heads.

Hence the current passes through the load in one direction only. It is unidirectional, and as shown in Fig. 2.15(b) it develops a fluctuating load voltage.

Notice from the above that the diodes conduct in pairs and that at any instant there are two diodes and the load across the secondary voltage. Thus the peak reverse voltage that each diode has to withstand when non-conducting is one-half of the voltage that the diodes in the previous circuits had to withstand for a given load voltage. This means that the bridge circuit requires diodes with only half the voltage rating, which is a further advantage of this circuit.

2.9 Smoothing

In each of the rectifying circuits considered, the resulting load voltage waveform consisted of unidirectional half-cycles. To convert these fluctuations into smooth direct current, a smoothing or filtering circuit is necessary.

Consider the effect of connecting a capacitor with a high value of capacitance (say 100 μF) across the load R_L of the circuit shown in Fig. 2.14. The diode voltage and current waveforms together with the resulting load voltage waveform are shown in Fig. 2.16. When considering the smoothing action of the circuit, assume that the first half cycle of the waveforms shown in Fig. 2.16 is due to diode D1 and the second half cycle is due to D2.

It may be seen that the rising anode voltage of D1 does not exceed the falling output voltage, which is also the cathode voltage, until time t_1. At this instant D1 conducts and current is supplied to the load and also to charge the capacitor. This continues until time t_2, at which instant D1 anode voltage falls below the cathode, whose voltage is constrained to follow the discharge voltage of C; therefore D1 becomes reverse biased and stops conducting. Current is now supplied to the load by the capacitor discharging through it until time t_3. During the time interval t_3 to t_4, D2 will be forward biased and so will pass the load current and also that necessary to replace the charge the capacitor lost during its discharge period t_2 to t_3.

Thus it may be seen that current is now supplied to the load alternatively by the transformer and the capacitor, i.e. the capacitor acts like a reservoir. It stores energy when the diodes conduct and releases it or discharges through the load in between diode conduction—hence it is called a reservoir capacitor. Also, if the time of discharge of the capacitor is low (i.e. the time constant *CR* is large) compared with the periodic time of the a.c. supply, the actual voltage across the load is much smoother.

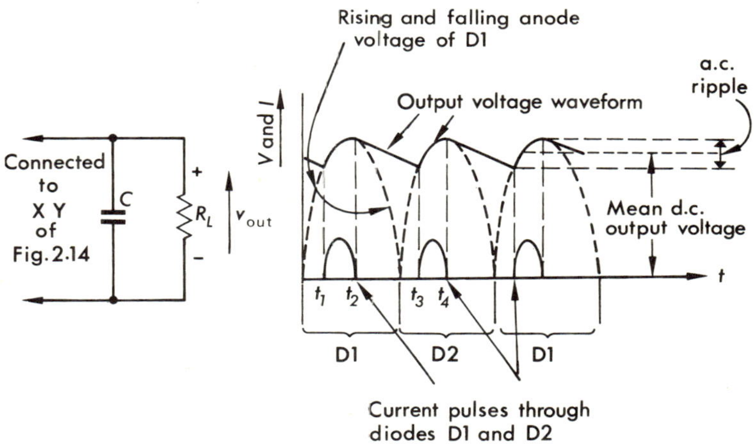

Fig. 2.16 Simple smoothing circuit and associated waveforms.

A further improvement on the output voltage may be obtained by inserting an a.c./d.c. separating circuit (Fig. 2.17) between the capacitor and the load. By considering the output waveform of the circuit shown in Fig. 2.16 it may be seen to consist of a large d.c. component, with a small a.c. ripple superimposed on it. A choke, which is a large inductor (5–20 H), is connected into the circuit of Fig. 2.17 such that the load current has to flow through it. Now, because the d.c. resistance of the choke is small compared with its reactance, the d.c. component of the load current will readily pass through. But to the a.c. component it will appear as a high

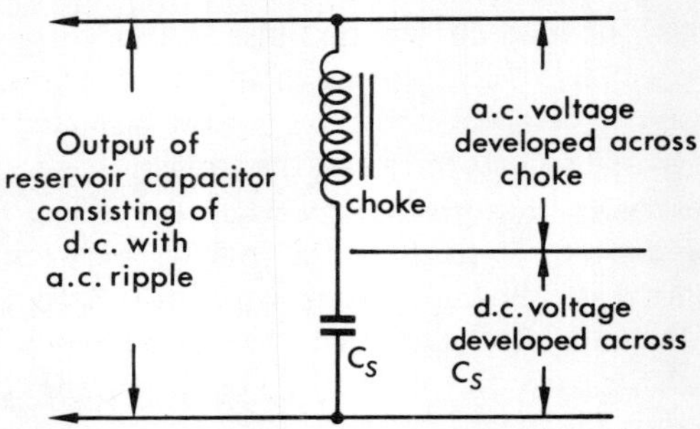

Fig. 2.17 *LC* filter.

blocking reactance and most of the ripple voltage will be developed across the choke. Conversely, if the value of C_S is large (say 100 μF), the a.c. component will 'see' almost a short-circuit, whilst to the d.c. component it will appear as a d.c. 'block'. Hence what remains of the a.c. ripple is by-passed to earth, while the d.c. voltage is developed across the capacitor.

Hence if the load is connected across C_S, as shown in Fig. 2.18, the resulting output voltage is relatively smooth.

2.10 Application of a Zener Diode as a Voltage Regulator

A voltage regulator is a device which maintains a constant voltage across a load, in spite of variations in either the load current or the supply voltage or both. The operating characteristic of the Zener diode, which is shown in Fig. 2.19, makes it an ideal control element for voltage regulation.

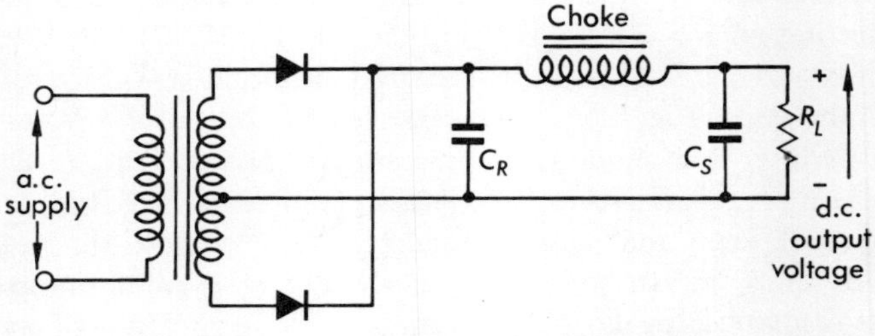

Fig. 2.18 Full-wave rectification and smoothing circuit.

2.10.1 *Zener diode characteristic*

A typical characteristic for a Zener diode is shown in Fig. 2.19. The breakdown voltage, which depends upon the manufacturing process, may be given any value between 3–200 V. The actual breakdown process is attributed to Zener breakdown, for diodes whose turnover voltage is less than 6 V. Above 6 V the breakdown is due to the avalanche or electron multiplication effect. Naturally there will be a small region of overlap where the process of breakdown is caused by both Zener and the avalanche effect. However, in general all semiconductor reference diodes which operate in the breakdown region of their characteristic are referred to as Zener diodes.

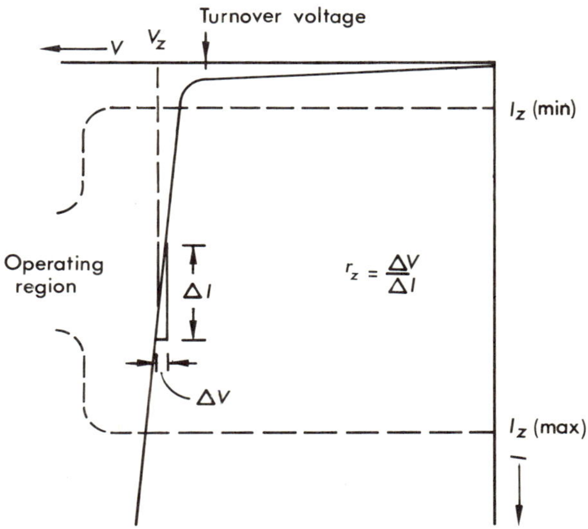

Fig. 2.19 Zener diode characteristic.

Due to the curvature of the characterìstic near breakdown, manufacturers always quote a minimum value of diode current, I_z (min). This is always greater than that required to avoid the curvature. Maximum diode current, I_z (max), and maximum power dissipation (*P* (max) for a given temperature) are also quoted.

D

Obviously, neither of these figures must be exceeded, otherwise the diode may suffer permanent damage.

The incremental slope resistance of the diode is also important. As shown in Fig. 2.19, this is simply $r_z = \Delta V/\Delta I$, taken anywhere over the operating region of the characteristic.

2.10.2 *Simple voltage regulation circuit*

Without regulation, the voltage across a load may vary due to:

(a) The effective resistance of the load changing, which will result in the load current changing.

(b) The supply voltage changing.

The function of the voltage regulator, or stabilizer as it is sometimes called, is to maintain a constant load voltage despite these effects.

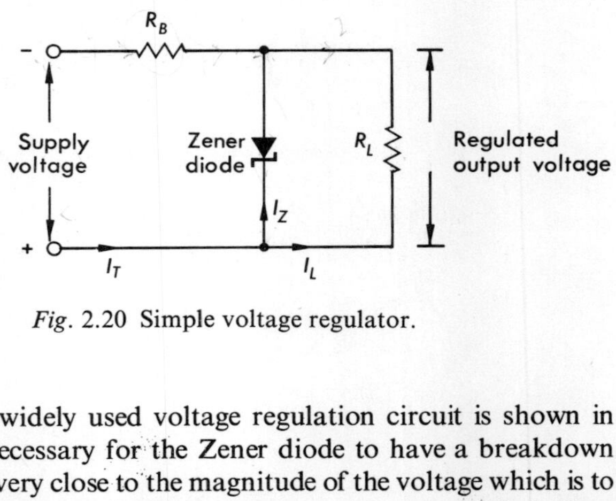

Fig. 2.20 Simple voltage regulator.

A simple but widely used voltage regulation circuit is shown in Fig. 2.20. It is necessary for the Zener diode to have a breakdown voltage which is very close to the magnitude of the voltage which is to be regulated. The current limiting resistor R_B is known as the ballast resistor and its value is chosen to give the correct operating point on the characteristic. Note that the ballast resistor and the diode form a potential divider across the supply voltage. The voltage dropped across the ballast resistor is proportional to the total current drawn from the supply (I_T). Also, the diode is connected in the reverse direction so that its anode is negative and its cathode positive.

To consider the action of the circuit, assume that the supply voltage increases by ΔV_S and that the load resistance R_L remains constant. The total current taken by the circuit will increase, the voltage ($I_T R_B$) dropped across R_B will increase, so that the actual increase of voltage across the diode and load, say δV_S, will be less than ΔV_S. The characteristic of the diode is such that the increase in voltage δV_S across it will cause its current to increase. Hence, the extra circuit current is diverted through the diode and not through R_L, and consequently the output voltage V_{out} will tend to remain constant.

If the supply voltage had decreased rather than increased the supply current would have decreased, causing the voltage drop across R_B to fall. The corresponding decrease in voltage across the diode would have resulted in its taking less current. Hence although I_T would have decreased, the larger decrease in diode current I_z would tend to offset it, thereby keeping the load current and output voltage constant.

Consider now a change in load current, the supply voltage remaining constant. If the load current increases, more current is drawn from the supply, and the voltage drop across R_B increases, tending to reduce the voltage drop across the diode. As a result current is diverted from the diode to the load, so tending to keep the load voltage constant. In the limit, when the load current I_L is a maximum I_z will be at a minimum approaching I_z(min). Conversely if the load current decreases the diode will compensate by taking more current so that once again the voltage across the load tends to remain constant.

REFERENCES

HUNTER, L. P. *Handbook of Semiconductor Electronics*, McGraw-Hill Book Company Inc., Section 3, 1962.

LECK, J. H. *Theory of Semiconductor Junction Devices,* Pergamon Press Ltd., 1967.

LEDGER, A. S. P. and N. H. ROCHE. *Fundamentals Electronics*, Blackie and Son Ltd., 1967.

MOODY, N. F. *Semiconductors and their Circuits*, The English Universities Press Ltd., Volume 1, 1966.

SHOCKLEY, W. 'The Theory of P–N Junctions in Semiconductors and P–N Junction Transistors', *Bell Systems Tech. J.*, vol. 28, 1949.

Chapter 3

Bipolar Transistors

3.1 Introduction

Fundamentally a bipolar transistor consists of two back-to-back pn junctions contained in a single semiconductor crystal. If the theory related to the pn junction given in the previous chapter is read carefully, then the principles of the transistor will be found relatively easy to understand.

As shown in Fig. 3.1, the two pn junctions give rise to three regions known as the **emitter**, **base** and **collector**. Transistors may be divided into two main categories depending on the order in which the doped semiconductor is arranged:

(a) npn, or (b) pnp.

The action of both is very similar, the essential differences being that in the pnp type the main current is carried by holes, whereas for the npn type the main current is carried by electrons. Because of this difference, the polarities of the biasing voltages are different for each type.

3.2 Operation of a Transistor

In order to simplify the understanding of transistors the following explanation is broken down into four easy-to-follow sections:

(a) Biasing
(b) Action
(c) Transistor currents
(d) Amplification

The npn transistor will be considered, but the theory developed is just as relevant for the pnp transistor, bearing in mind the different carriers and polarity of biasing. A brief examination of the pnp transistor will be made later in this chapter.

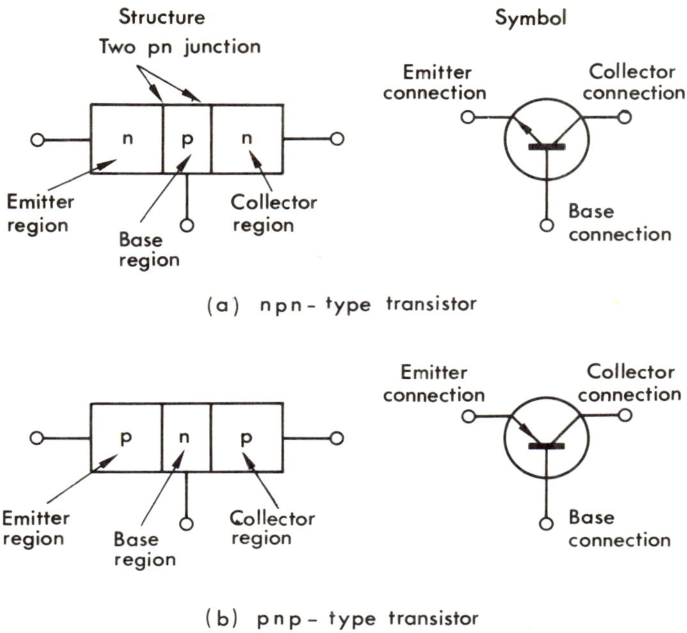

Fig. 3.1 Basic structure and symbols for a npn and pnp transistor.
(a) npn-type transistor.
(b) pnp-type transistor.

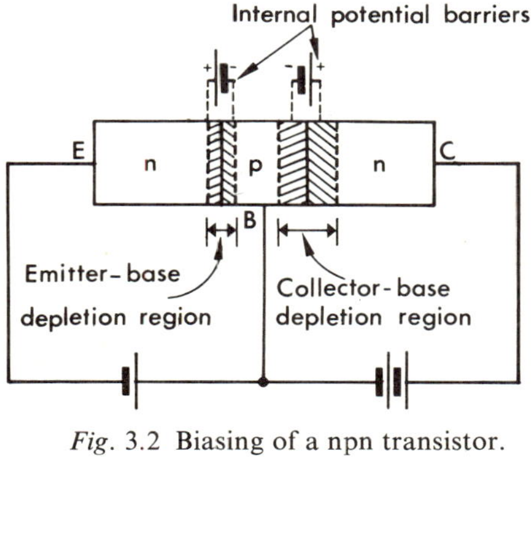

Fig. 3.2 Biasing of a npn transistor.

From Fig. 3.2 it may be seen that the transistor is in effect two semiconductor diodes placed back-to-back. At each junction there will be a depletion layer which forms an internal potential barrier. The polarities of the two potential barriers are as indicated.

External biasing is provided by batteries B1 and B2. B1 is connected so as to neutralize the internal potential barrier across the emitter–base junction, whilst B2 is connected so as to aid the internal potential barrier across the collector–base junction. Thus the first junction, the **emitter–base junction, is forward biased,** whilst the **collector–base junction is reverse biased**. This must always be so, in order to achieve amplification with transistors.

3.2.1 *Transistor action*

When the negative potential of B1 in Fig. 3.3 is applied to the emitter, the negatively-charged electrons are repelled towards the junction. Since the emitter–base potential barrier has been neutralized, they flow into the base region, which is made very thin, say 10 μm, and they quickly come under the influence of the reverse bias across the collector–base junction. Because this junction is reverse biased, the majority carriers (holes) in the base are blocked. But most of the electrons, which have come from the emitter, diffuse across the base region and successfully enter the collector region.

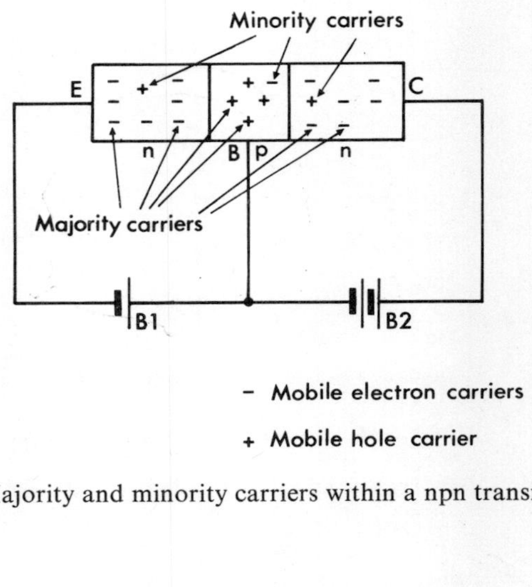

Fig. 3.3 Majority and minority carriers within a npn transistor.

Once the electrons are in the collector they come under the attractive influence of the positive terminal of the battery B2.

Although the base is made very thin, the injected electrons take a small but finite time to penetrate this region before coming under the influence of the collector–base reverse bias. During this time interval some of the electrons meet with holes (which are the majority carriers in the base) and recombine. In order to keep this recombination to a minimum, the base is not only made very thin, but its level of doping is made less than the emitter's. There are then fewer holes and so the probability of an electron meeting a hole is further reduced. In practice about 98 per cent of the electrons which are injected into the base from the emitter successfully reach the collector.

3.2.2 *Current flow in a npn transistor*

Just as in the diode, the number of current carriers within a n- or p-type region always tends to remain constant. This equilibrium is maintained as follows: for every electron which is injected into the base region from the emitter, another one is supplied to the emitter by an electron leaving the negative terminal of B1 in Fig. 3.4 and flowing into the emitter region. The flow of such electrons constitutes the emitter current I_E of the transistor, and its magnitude is dependent on the voltage of battery B1. For every hole lost in the base region due to recombination another is created by an electron leaving the base and flowing into the positive terminal of B1. Such electrons

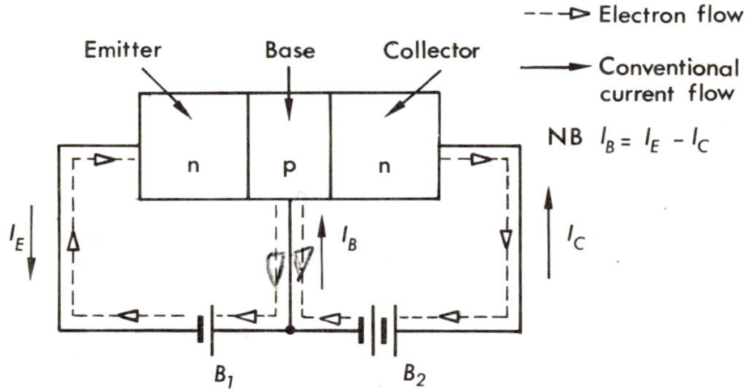

Fig. 3.4 Current flow in a npn transistor

form the base current I_B. Finally, the collector current I_C is formed by the electrons which reach the collector from the emitter and are attracted to the positive terminal of B2.

Thus the emitter current is dependent on the magnitude of the forward bias across the emitter–base junction, and the base and collector currents are dependent on the emitter current, i.e. $I_E = I_B + I_C$. The fraction of the emitter current which reaches the collector is represented by the symbol α (alpha) so that

$$\begin{aligned}\text{Emitter current:} \quad & I_E = I_B + I_C .\\ \text{Collector current:} \quad & I_C = \alpha I_E .\\ \text{Base current:} \quad & I_B = I_E - I_C \\ & = I_E - \alpha I_E .\\ & = I_E(1-\alpha) .\end{aligned}$$

A typical value for α is 0·98, therefore for an emitter current of 1 mA, $I_C = 0{\cdot}98$ mA and $I_B = 0{\cdot}02$ mA.

3.2.3 *Amplification*

The word transistor is derived from the words **transfer resistor.** This is not surprising, since this virtually describes how amplification in a transistor is achieved. Consider Fig. 3.5 which shows the

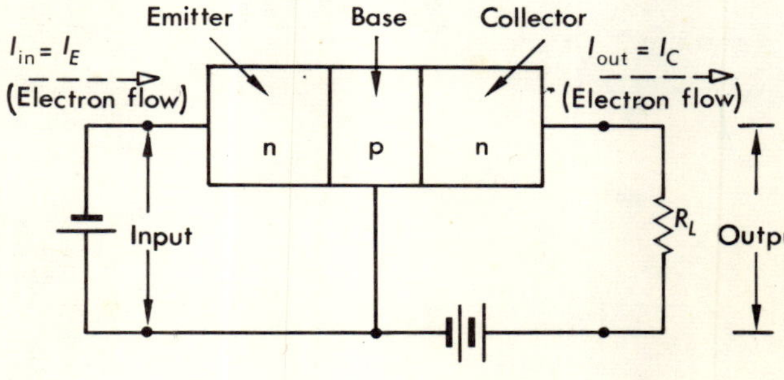

Fig. 3.5 d.c. biased transistor.

basic circuit of a d.c. biased transistor. Although, in general, alternating signals will be of prime importance when considering transistor amplification, it is possible to understand how amplification is achieved by considering steady input and output currents and voltages.

Let the d.c. forward bias produce a steady input or emitter current (I_{in} or I_E) which gives rise to a steady output or collector current (I_{out} or I_C). Since α is almost unity, the input and output currents of the transistor are approximately equal. *But* the input and output resistances of the transistor are enormously different on account of biasing. The forward biasing of the emitter–base junction gives an input resistance R_{in} of say, 50 Ω, whereas the reverse biasing of the collector–base junction gives an output resistance of, say, 500 kΩ.

Because of this high collector-to-base resistance a load resistor R_L can therefore be added in series without changing the value of I_{out} very much. For example consider R_L to be 5 k Ω. The output voltage V_{out} is equal to the product of $I_{out} \, . \, R_L$. Hence

$$\text{voltage gain} = \frac{V_{out}}{V_{in}} = \frac{I_{out} \, . \, R_L}{I_{in} \, . \, R_{in}}$$

$$\simeq \frac{R_L}{R_{in}}, \qquad \text{since } I_{out} \simeq I_{in},$$

$$\simeq \frac{5000}{50} = 100.$$

Thus it may be seen that the transistor achieves a large voltage gain on account of its input and output resistances being of different magnitudes, whilst its input and output currents are approximately equal.

3.2.4 *Operation of a pnp transistor*

At the beginning of this chapter, it was pointed out that bipolar transistors may be broadly divided into two main categories: npn or pnp. The operation of the npn type has been considered in some detail. In order to show that the operation of both types is very similar, a brief examination of the pnp type shown in Fig. 3.6 will now be made.

The doping of the emitter, base and collector is respectively opposite to that of a npn transistor. That is, the main current is now carried by holes instead of electrons. This necessitates the reversal of the polarities of the external biasing, which is provided by batteries B1 and B2. The emitter–base junction is still forward biased but, in order to neutralize the internal potential barrier at the junction and at the same time inject holes into the base, the emitter must be positive with respect to the base.

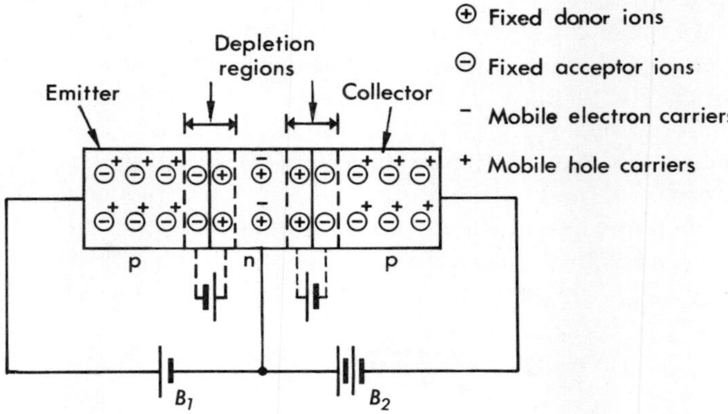

Fig. 3.6 pnp transistor.

The holes, on entering the base, diffuse or drift across this relatively thin region. Although a few are lost due to recombination with electrons, the majority of them successfully reach the collector–base junction where, owing to the action of the reverse biasing, they are attracted across the junction and enter the collector region. Once in this region they are attracted through until they are neutralized by electrons from the negative terminal of B2.

3.2.5 *Current flow in a pnp transistor*

The direction of the currents may be deduced by considering the movement of carriers within the system. For every hole which is injected into the base from the emitter, another hole is created in the emitter by an electron leaving the emitter region and moving to the positive terminal of B1. The flow of such electrons constitutes the emitter current I_E of the transistor. Also, for every electron lost in the base region due to recombination, another electron is supplied to the base by the negative terminal of B1. Such electrons form the base current I_B. Finally, the collector current I_C is formed by by electrons leaving the negative terminal of B2 and recombining with holes which have reached the collector from the emitter region.

Thus, due to the different type of current carrier and opposite polarity biasing, the currents in a pnp transistor flow in the opposite direction compared with those in a npn transistor.

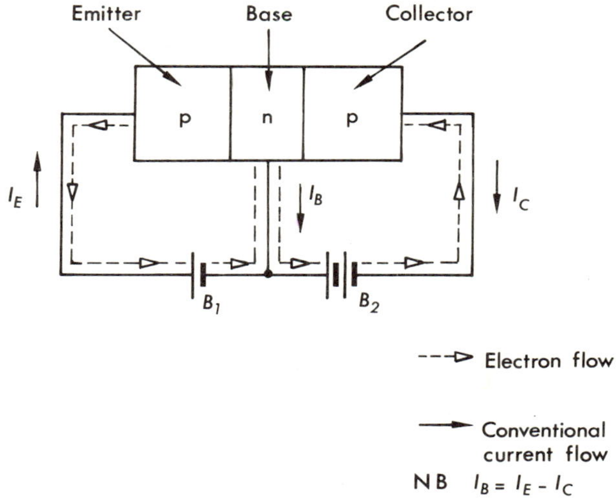

Fig. 3.7 Current flow in a pnp transistor.

Manufacturers will supply both npn and pnp transistors having identical characteristics and ratings. Consequently an engineer may choose the more appropriate type for particular requirements.

3.3 Transistors in Practice

So far in this chapter, the mechanism of operation of the transistor has been examined. Important points to remember are that:

(a) The emitter–base junction is forward biased, giving a relatively low resistance.
(b) The collector–base junction is reverse biased, giving a large resistance.
(c) The emitter and collector currents are approximately equal, the small difference being equal to the base current.

(d) Amplification is obtainable by virtue of the difference in output and input resistance, with the respective currents being approximately equal.

When the transistor is used in practice, there are three possible ways in which it may be connected into a circuit. The input is connected between one pair of terminals and the output is taken from another pair. Since the transistor has only three terminals, it follows that one terminal must always be common to both input and output. This is illustrated in Fig. 3.8.

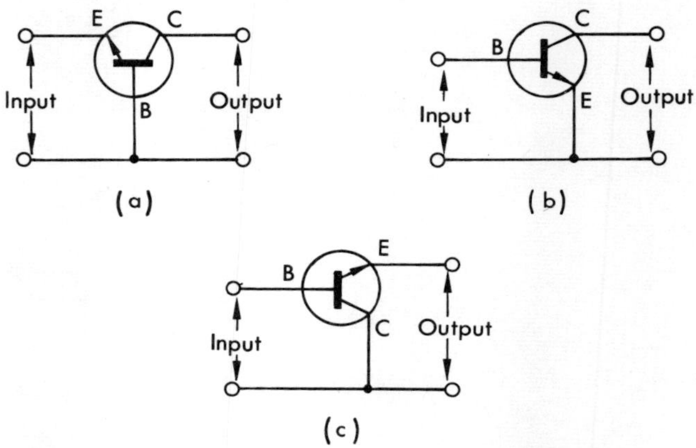

Fig. 3.8 The three possible connections for a transistor.
(a) Common-base (CB).
(b) Common-emitter (CE).
(c) Common-collector (CC).

Each of the connections or configurations shown in Fig. 3.8 has its own particular properties which dictate the way in which the transistor is connected in practice. A number of these properties, together with the relationship between the currents and voltages of the various terminals, may be illustrated by three characteristic curves. These are:

(a) the input characteristic,
(b) the output characteristic,
(c) the transfer characteristic.

Each of the above connections, together with their associated characteristic curves will now be considered.

3.3.1 *Common-base (CB) connection and characteristics*

With the common-base connection, shown in Fig. 3.9(a), the input is applied between the emitter and base and the output is taken from between the collector and base. The base is the common terminal—hence its name. The d.c., or steady, currents and voltages are as indicated. Note that in transistor terminology, d.c. currents and voltages are denoted by uppercase (or capital letters), whereas a.c. currents and voltages are denoted by lower case letters. Subscripts are used to identify the currents and voltages associated with the different transistor terminals. For example, V_{EB} represents the d.c. voltage between the emitter and base terminals, whereas v_{eb} would represent the a.c. voltage between the same two terminals.

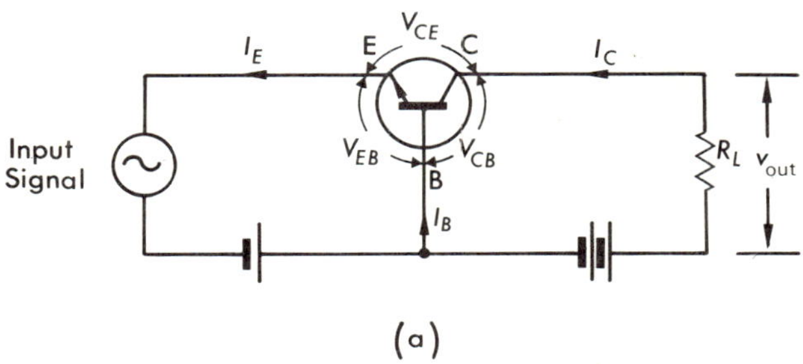

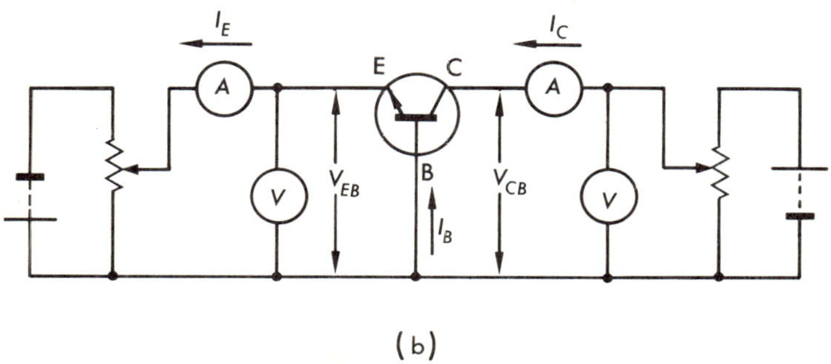

Fig. 3.9 The transistor connected in the CB configuration.
(a) Basic CB connection.
(b) Suitable circuit diagram for obtaining static characteristic curves.

The static characteristics of a transistor connected in the common-base configuration may be determined by the use of the circuit diagram shown in Fig. 3.9(b).

3.3.2 *Input characteristic for the common-base connection*

This characteristic may be easily obtained by varying the emitter-base bias V_{EB} and measuring the resulting emitter current I_E with the collector–base voltage V_{CB} maintained constant throughout. Figure 3.10 shows a typical input characteristic for a low-power transistor.

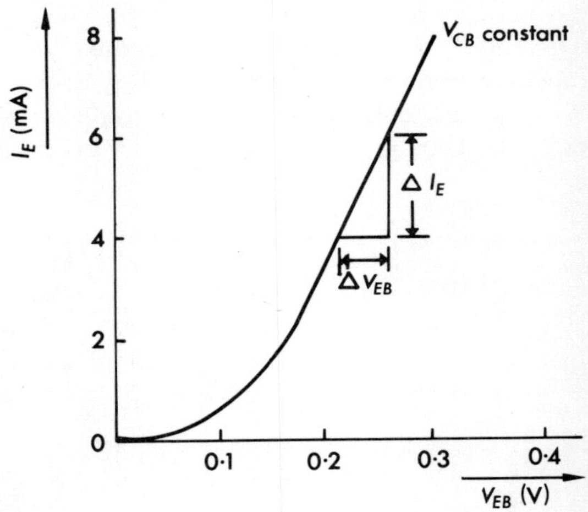

Fig. 3.10 Input characteristic for CB connection.

The overall shape of the characteristic may be seen to be the forward characteristic of a p–n diode which, in essence, is what the emitter base junction is. The input resistance of the transistor to a.c. is given by the reciprocal of the slope, i.e.

$$R_{in} = \frac{\Delta V_{EB}}{\Delta I_E}.$$

It is important to note that the input resistance of the transistor is initially nonlinear. A typical value for R_{in} over the linear part of the characteristic is 50 Ω, but for low values of V_{EB} it is considerably greater. This is an inherent disadvantage of the transistor and unless care is taken it may lead to distortion of the amplified signal.

3.3.3 *Output characteristic for the common-base connection*

The output characteristic is obtained by adjusting V_{EB} to give a certain value of I_E and then plotting variations of I_C against V_{CB}.

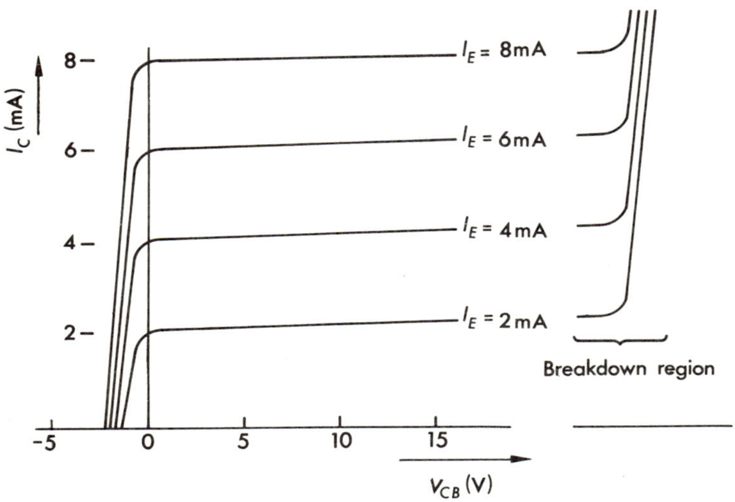

Fig. 3.11 Output characteristic for CB connection.

To obtain a family of such curves the above procedure is repeated for various values of I_E. A typical output characteristic is shown in Fig. 3.11 from which the following important points may be noted:

(a) The output resistance of the transistor R_{out}, which is given by the reciprocal of the slope of the near horizontal part of the characteristic, will be very large. This must be so, since I_C is virtually independent of V_{CB}. $R_{out} = \Delta V_{CB}/\Delta I_C$ and a typical value is 500 kΩ.

(b) It may be seen that not only does I_C fail to increase with increase in V_{CB} (below breakdown) but that I_C flows even when V_{CB} is zero. This is because electrons are being injected into the base region under the action of the forward-biased emitter–base junction and are being collected by the collector due to the action of the internal potential barrier at the collector–base junction. In order to reduce I_C to zero, it is necessary to neutralize this internal potential by applying a small forward bias across the collector–base junction.

(c) An estimation of the common-base forward current gain h_{fb} may be made. h_{fb} is defined as the ratio of the change of I_C to the corresponding change of I_E with V_{CB} maintained constant. This is usually expressed as

$$h_{fb} = \frac{\Delta I_C}{\Delta I_E} \quad (V_{CB} \text{ constant}).$$

The symbol α is also used for the common-base current gain, i.e.

$$\alpha = |h_{fb}|.$$

A typical value for $|h_{fb}|$ is 0·98. (The | | sign indicates the the arithmetical value).

(d) I_C depends upon I_E. The small difference is due to electrons being lost in the base owing to recombination.

(e) Although I_C is virtually independent of V_{CB} over the working range of the transistor, if V_{CB} is permitted to increase, then eventually I_C increases rapidly due to avalanche breakdown.

3.3.4 *Transfer characteristic for the common-base connection*

The transfer characteristic shown in Fig. 3.12 illustrates the typical relationship between input current I_E and output current I_C with the output voltage V_{CB} held constant. Its slope $\Delta I_C/\Delta I_E$ is the forward current gain h_{fb}.

Fig. 3.12 also shows an enlarged view of the characteristic at the origin. It may be noticed that I_C is not zero when I_E is zero. The small current I_{CBO}, is the transistor's leakage current due to minority carriers flowing across the reverse biased collector–base junction. (The symbol I_{CBO} is used to indicate that the leakage current I flows between collector C and base B with emitter current zero, i.e. I_{CBO}).

For a low-power germanium transistor, the leakage current at room temperature, 20°C, is typically 5 μA, whereas for a similar silicon transistor it is less than one thousandth of this. This is one of the chief reasons why most present-day transistors are made from silicon rather than germanium. Although a leakage current of

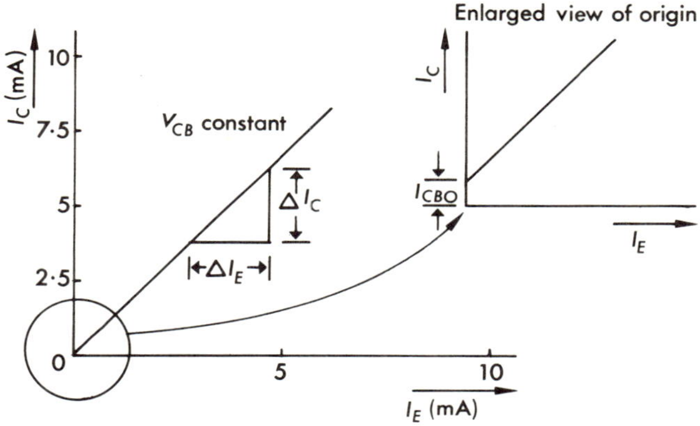

Fig. 3.12 Transfer characteristic for CB connection.

5 μA at 20°C does not seem very large, it becomes very serious at higher temperature because of the temperature dependence of I_{CBO}. For example, it approximately doubles for about every 8°C rise in temperature. Furthermore, when the transistor is connected in the common-emitter configuration the leakage current is amplified due to transistor action, so that 5 μA may become 250 μA. In fact, it is this leakage current which imposes a maximum operating temperature of approximately 80°C for germanium and approximately 160°C for silicon transistors.

3.4 Simple Common-base Amplifier

The currents shown in Fig. 3.13 are d.c. and are caused by the two biasing voltages V_1 and V_2. If a sinusoidal input voltage is applied it will be superimposed on the existing d.c. voltage V_1. Thus as the input signal voltage increases in a positive direction, the effective forward bias between emitter and base will decrease, so causing I_E and consequently I_C to decrease, which in turn will decrease the voltage drop across the load resistor R_L ($V_{\text{Load}} = I_C R_L$). Therefore, since the output voltage is given by $V_2 - I_C R_L$ the output voltage must become more positive (see output voltage waveform, Fig. 3.13).

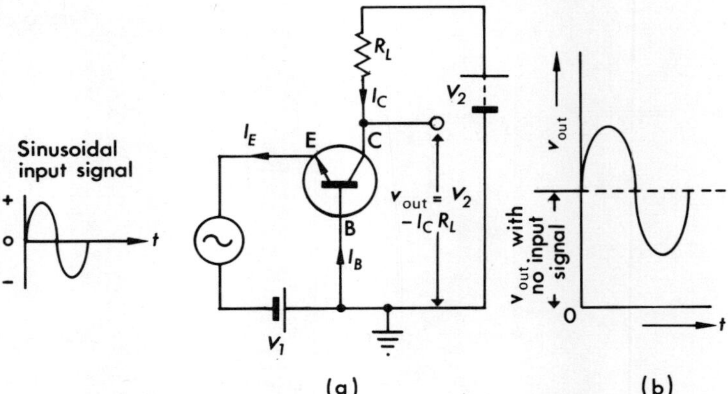

Fig. 3.13 Simple CB amplifier.
(a) Circuit.
(b) Output voltage waveform.

For the negative half-cycle of the input signal voltage the forward bias increases. Thus I_E increases, producing a corresponding increase in I_C and a corresponding increase in the voltage drop across R_L. As a result the output voltage goes less positive. (See output voltage waveform, Fig. 3.13.)

Hence it can be seen that the output voltage rises and falls in sympathy with the input signal voltages. It is said that the output and input voltages are in phase or, alternatively, that the common-base amplifier introduces no phase shift between input and output signals.

3.4.1 *Voltage and power amplification*

In order to calculate the voltage and power amplification or gain of an amplifier, it is necessary to know some details of the transistor and the amplifier circuit.

Assume that the transistor has input and output resistances of 50 Ω and 500 kΩ respectively and a forward current gain h_{fb} of 0·98. Let the load resistor R_L connected in the collector lead be 5 kΩ.

If the input signal voltage v_s is such that the resulting alternating input current i_e is 1 mA, then by Ohm's law

$$v_s = i_e \, . \, R_{\text{in}} = 1 \cdot 10^{-3} \times 50 = 0{\cdot}05 \text{ V}.$$

The corresponding alternating output current i_c will be 0·98 mA ($i_c = h_{fb} \, . \, i_e$), so that the resulting change in the output signal v_{out} developed across R_L will be

$$v_{\text{out}} = i_C R_L = 0{\cdot}98 \times 10^{-3} \times 5 \times 10^3 = 4{\cdot}9 \text{ V}.$$

Hence

$$\text{voltage gain} = \frac{v_{\text{out}}}{v_s} = \frac{4{\cdot}9}{0{\cdot}05} = 98.$$

The power gain of an amplifier is given by the ratio Output power/Input power, or alternatively since power in a resistor = current × voltage,

$$\text{power gain} = \text{current gain} \times \text{voltage gain}.$$

Therefore for the amplifier considered,

$$\text{power gain} = 0{\cdot}98 \times 98 = 96{\cdot}04.$$

3.4.2 *Disadvantages and advantages of the common-base amplifier*

Although the CB connection may be the easiest configuration in which to explain the operation of the transistor, it is not the most widely used on account of the following disadvantages:

(a) The input resistance is so low that a very low source resistance is necessary to drive it and such a source resistance is comparatively rare.

(b) The considerable difference between input and output resistances (50 Ω and 500 kΩ) makes matching between cascaded stages difficult, unless expensive step-down matching transformers are used.

(c) Although the CB connection has a high voltage gain, its current gain is always less than unity and hence the power gain is restricted.

The major disadvantages listed above are sometimes offset in certain applications due to the following advantages of the CB connection:

(a) Leakage current is less troublesome than in the other two connections.
(b) The transistor will operate at a much higher frequency than in the CE connection.
(c) A transistor connected in CB is not so sensitive to variations in transistor parameters as it is in the CE connection.

3.5 Common-emitter (CE) Connection and Characteristics

Fig. 3.14 (a) shows the common-emitter connection, which is the most widely used of the three. The input is applied between base and

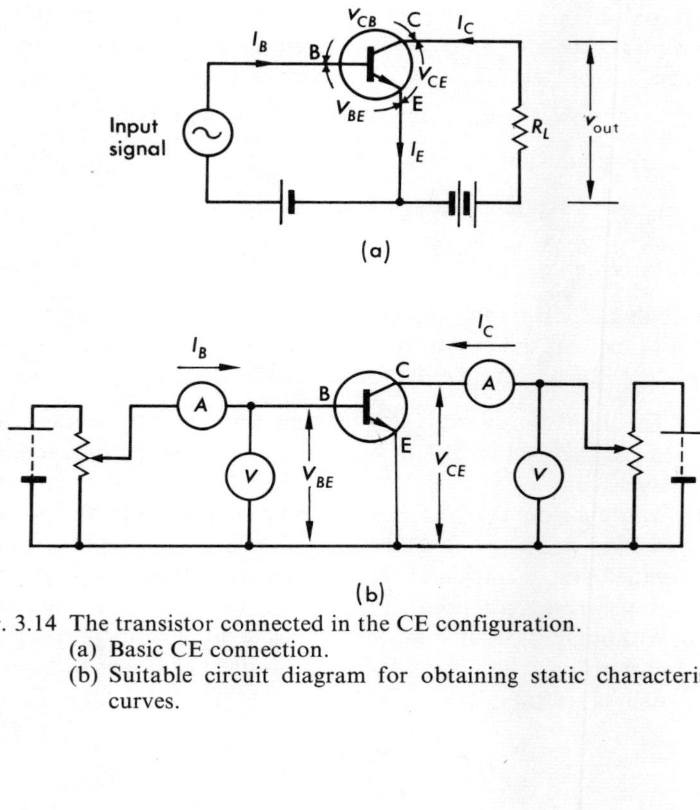

Fig. 3.14 The transistor connected in the CE configuration.
(a) Basic CE connection.
(b) Suitable circuit diagram for obtaining static characteristic curves.

emitter and the output is taken from between the collector and emitter—the emitter is the common connection.

The static characteristic for a transistor connected in the CE configuration may be determined by the use of the circuit diagram shown in Fig. 3.14 (b). The procedure is similar to that outlined for the CB connection.

3.5.1 *Input characteristic for the common–emitter*

The input characteristic shown in Fig. 3.15 illustrates the way in which the input current I_B varies with input voltage V_{BE} for a

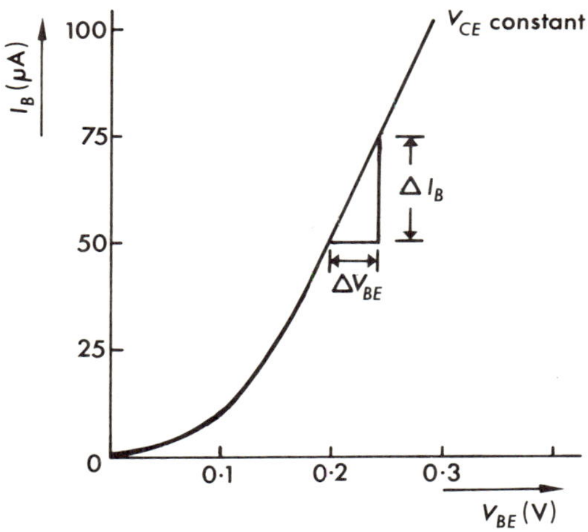

Fig. 3.15 Input characteristic for CE connection.

constant collector–emitter voltage V_{CE}. Like the CB connection, the overall shape resembles the forward characteristic of a p–n diode. Due to the initial nonlinearity between I_B and V_{BE}, the input resistance varies considerably. For example, the reciprocal of the slope at $I_B = 10\ \mu A$ gives an input resistance R_{in} of approximately 4 kΩ, whereas over the more linear part of the characteristic, say $I_B = 100\ \mu A$, R_{in} is approximately 600 Ω. Although the input resistance of the CE connection is considerably greater than that of the CB connection, it is still relatively low.

3.5.2 *Output characteristic for the common-emitter connection*

A typical output characteristic for a transistor connected in the common-emitter configuration is shown in Fig. 3.16. It illustrates

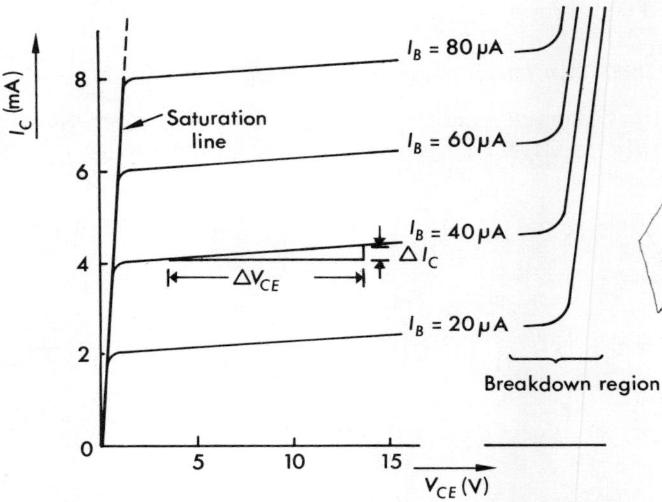

Fig. 3.16 Output characteristic for CE connection.

how the output current I_C varies with output voltage V_{CE} for a constant input current I_B. As V_{CE} increases from zero, I_C rapidly increases to a near saturation level for a fixed base current. This occurs since for a base current of, say, 20 μA the emitter–base junction must be forward biased. From section 3.2.3,

$$I_E = I_B/(1-\alpha) = I_B/1-|h_{fb}|.$$

If $|h_{fb}| = 0{\cdot}98$, then $I_E = 2$ mA. Hence as soon as V_{CE} increases positively, I_E increases sharply to 2 mA, resulting in I_C increasing similarly.

I_C levels as it approaches 2 mA but complete saturation does not occur because of two factors:

(a) As V_{CE} increases, the collector–base depletion layer increases in width, so causing the effective base width to decrease. This has the effect of increasing h_{fb}.

(b) As V_{CE} increases, slight avalanche multiplication occurs across the collector–base junction resulting in an increase in I_C.

Note that if V_{CE} is allowed to increase too far, the collector–base junction completely breaks down, and due to this avalanche breakdown I_C increases rapidly, possibly causing permanent damage to the transistor.

It may be seen on the output characteristic that all the near horizontal lines, representing different values of base current, merge into a single line at low values of V_{CE}, and this line is called the **transistor's saturation line.** Saturation occurs when the collector is unable to collect all the carriers injected into the base by the forward bias. In other words, saturation occurs when a change in I_B does not produce a corresponding change in I_C. The reciprocal of the slope of the saturation line gives the collector saturation resistance of the transistor. For a low-power transistor this may be typically 10–100 Ω, whereas for a large power device it may be less than 1 Ω.

When used as an amplifier, the transistor is so biased that it operates over the near-horizontal part of the characteristic. A typical value for the output resistance in this region (given by the ratio $\Delta V_{CE}/\Delta I_C$) is 10 kΩ to 50 kΩ—much lower than that for the CB connection which is approximately 500 kΩ.

3.5.3 *Transfer characteristic for the common-emitter connection*

Fig. 3.17 illustrates the typical relationship between output current I_C and input current I_B for a fixed value of V_{CE}. The slope $\Delta I_C/\Delta I_B$ is a measure of the forward current gain (symbol h_{fe} or β) for the transistor in the common-emitter mode. A typical value may be between 40 and 200. This, of course, is much greater than the current gain for the CB connection which is always less than unity.

Note 1: $|h_{fb}| = \Delta I_C/\Delta I_E$, and since the general current convention for transistors is that currents entering the transistor are considered positive and currents leaving are considered negative, it follows that ΔI_C and ΔI_E have different signs, so that h_{fb} will be negative, i.e. $\alpha = -h_{fb}$.

Note 2. Because $\Delta I_E = \Delta I_B + \Delta I_C$ and $\Delta I_C/\Delta I_E = -h_{fb}$, the relationship between h_{fe} and h_{fb} may be easily determined. Combining these expressions gives $\Delta I_B = \Delta I_E(1 + h_{fb})$. Substituting first for ΔI_B and then for $\Delta I_C/\Delta I_E$ we find

$$h_{fe} = \frac{\Delta I_C}{\Delta I_B}$$

$$= \frac{\Delta I_C}{\Delta I_E(1+h_{fb})}$$

$$= \frac{\Delta I_C}{\Delta I_E} \cdot \frac{1}{(1+h_{fb})}$$

$$= \frac{-h_{fb}}{1+h_{fb}}$$

For example, if $h_{fb} = 0{\cdot}98$, then

$$h_{fe} = \frac{0{\cdot}98}{1 - -0{\cdot}98} = 49.$$

From Fig. 3.17 it may be seen that a small collector current flows even though I_B is zero. This is the common-emitter leakage current and it is given the symbol I_{CEO} because the leakage current flow between collector C and emitter E with base current zero. Although this leakage current has the same source as I_{CBO}, namely minority carriers crossing the reverse-biased collector–base junction, it is much larger. This increase in leakage current, which is caused by transistor amplification, may be explained as follows. Consider the circuit shown in Fig. 3.18. Here the base of the transistor has been left open-circuit and therefore the base current must be zero ($I_B = 0$).

Let I_{CBO} be the leakage current flowing across the reverse-biased collector–base junction. Since $I_B = 0$, this leakage current I_{CBO} must be balanced or cancelled by an equal and opposite current from the base. The fraction of the emitter current which flows into the base will be $I_E(1+h_{fb})$ (see section 3.2.2, where $-h_{fb}$ is substi-

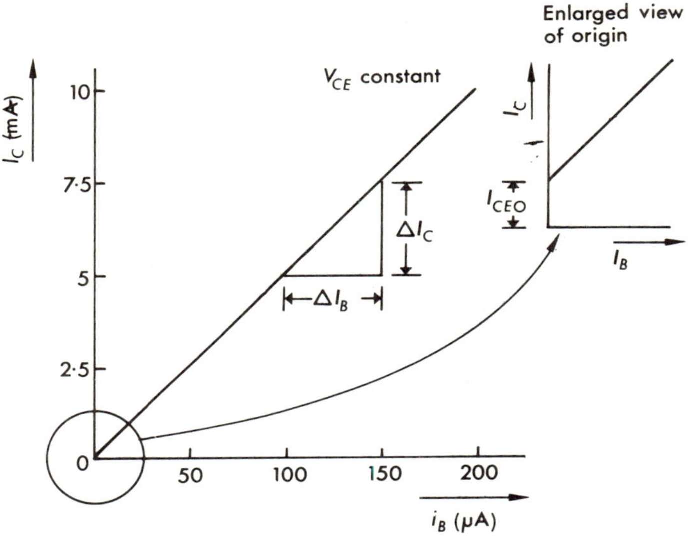

Fig. 3.17 Transfer characteristic for CE connection.

tuted for α). Therefore if $I_B = 0$,

$$I_E(1+h_{fb}) = I_{CBO},$$

$$I_E = \frac{I_{CBO}}{(1+h_{fb})}.$$

This is the current which must flow out of the emitter so that a fraction of it may flow into the base to offset I_{CBO}. What remains

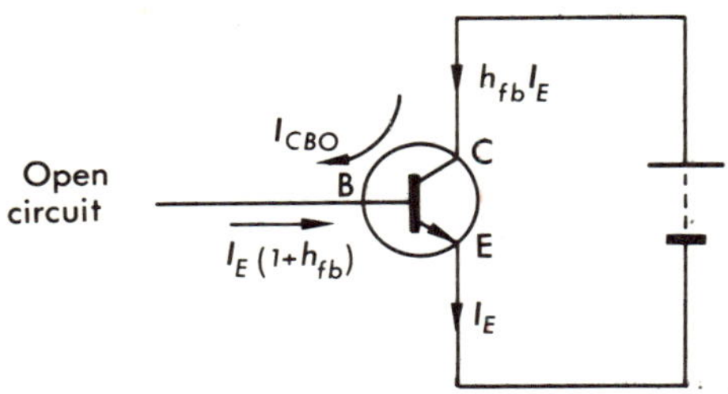

Fig. 3.18 Leakage current I_{CEO}.

of the emitter current flows into the collector and since $I_C = -h_{fb} I_B$, the fraction which reaches the collector will be

$$\frac{-h_{fb} I_{CBO}}{1+h_{fb}}$$

Hence the total collector current will have two components:

(a) I_{CBO} which flows directly from collector to base,
(b) the fraction of the emitter current, $-h_{fb}/(1+h_{fb}) . I_{CBO}$ or, using the equation derived in Note 2 above, $h_{fe} . I_{CBO}$.

Therefore

$$\begin{aligned} I_C &= I_{CBO}+h_{fe} I_{CBO} \\ &= I_{CBO}(1+h_{fe}) \end{aligned}$$

This is the total collector leakage current of the CE connection and is represented by I_{CEO}.

For a low-power germanium transistor at 25°C, I_{CBO} is typically 5 μA, which makes I_{CEO} about 250 μA, if $h_{fe} = 49$. For a similar silicon transistor at the same temperature, I_{CBO} is typically only 1 nA, which gives an I_{CEO} of only 50 nA. Because this leakage current I_{CEO} approximately doubles for every 10°C rise in temperature, particularly for transistors made from germanium and connected in the CE configuration, it is a very important parameter.

3.6 Simple Common-emitter Amplifier

Consider the amplifier shown in Fig. 3.19. When the alternating input voltage increases positively from zero, the effective forward bias between base and emitter is increased. I_E increases, causing a corresponding increase in I_C and in the voltage dropped across R_L. Therefore the collector potential must become less positive, i.e. the output voltage falls (see the waveform in Fig. 3.19). When the alternating input voltage increases in a negative direction, i.e. on the negative half-cycle, the effective forward bias is decreased, so I_E decreases causing I_C to follow. The corresponding voltage drop across R_L decreases, so that V_{out} rises positively (see waveform).

Thus, with the common-emitter connection when the input rises the output falls and *vice versa*. Consequently it is said that the common-emitter amplifier introduces phase reversal of the signal.

An example giving current, voltage and power gains of the common-emitter amplifier will be considered later in this chapter.

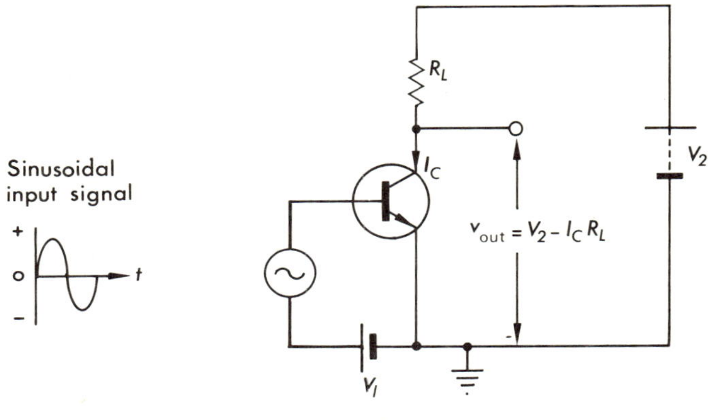

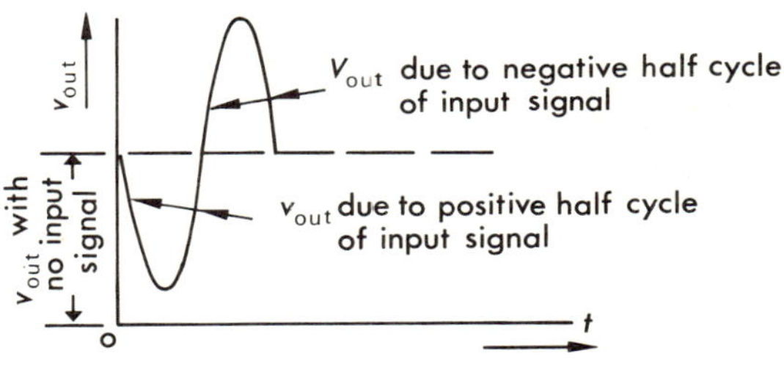

Fig. 3.19 Simple CE amplifier.
(a) Circuit.
(b) Output voltage waveform.

3.7 Advantages and Disadvantages of the Common-emitter Connection

The common-emitter connection is by far the most widely used configuration of the three, because of the following advantages:

(a) It has significant current and voltage gains, so therefore the highest power gain.
(b) The difference between its input and output resistances is not so extreme and this simplifies interconnection between stages.

The two major advantages listed above are generally considered more important than the following disadvantages:

(a) Leakage current may be troublesome at high temperatures.
(b) Frequency response is inferior to that of the other two connections, because of the larger effective input capacitance.

3.8 Common-collector (CC) Connection

With the common-collector configuration shown in Fig. 3.20, the input is applied between the base and the collector while the output is taken from between the emitter and collector. At first sight this may not appear to be so, but it must be remembered that the impedances of the bias batteries are virtually zero.

When the common-base and common-emitter connections were considered (sections 3.3 and 3.5), typical characteristic curves illustrating the relationship between current and voltage of the various terminals were examined. With the common-collector connection, the information supplied by such curves is not so useful and so the curves are seldom supplied by manufacturers. Consequently the special features of the common-collector connection will be considered without the help of characteristic curves.

Current gain. The forward current gain of the common collector connection, (symbol h_{fc}) which is defined as $\Delta I_E/\Delta I_B$ (with V_{CE} constant), is the highest of the three connections. For example,

$$h_{fe} = \frac{\Delta I_C}{\Delta I_B}$$

whence

$$\Delta I_C = h_{fe}\Delta I_B\,;$$

also

$$\Delta I_E = \Delta I_C + \Delta I_B\,.$$

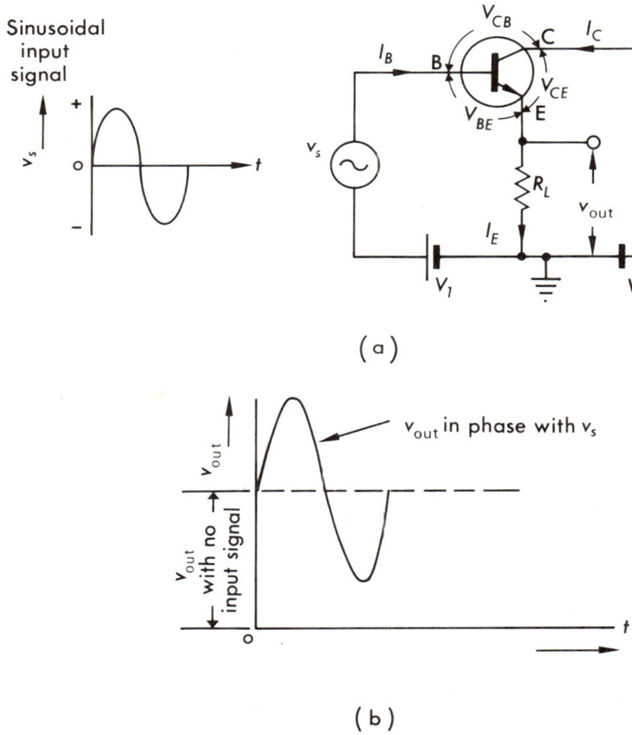

Fig. 3.20 Simple CC amplifier.
(a) Circuit.
(b) Output voltage waveform.

Thus, substituting for ΔI_E and ΔI_C in turn, we find

$$h_{fc} = \frac{-\Delta I_E}{\Delta I_B} \qquad (\Delta I_E, \Delta I_B \text{ have different signs})$$

$$= \frac{-(\Delta I_C + \Delta I_B)}{\Delta I_B}$$

$$= \frac{-(h_{fe}\Delta I_B + \Delta I_B)}{\Delta I_B}$$

$$= -(h_{fe}+1).$$

Emitter-follower. Fig. 3.20 shows that when the alternating input signal voltage increases from zero in a positive direction, the forward bias of the transistor increases. This causes I_E to increase, which causes a corresponding increase in the output voltage ($v_{out} = I_C R_L$). If the input voltage goes negative, the forward bias decreases, and I_E correspondingly decreases, causing v_{out} to follow. Thus the input and output voltages rise and fall together and so it may be said that the emitter potential follows the base potential. It is because of this that the common-collector connection is more generally referred to as the *emitter-follower*.

Voltage gain. It may be seen from Fig. 3.20 that since the emitter voltage follows the base voltage, the actual effective alternating voltage between the base and the emitter is equal to the difference between input and output voltages. In other words, the voltage (v_{eb}) which is amplified by the transistor is not the actual input signal, but $v_{eb} = v_s - v_{out}$.

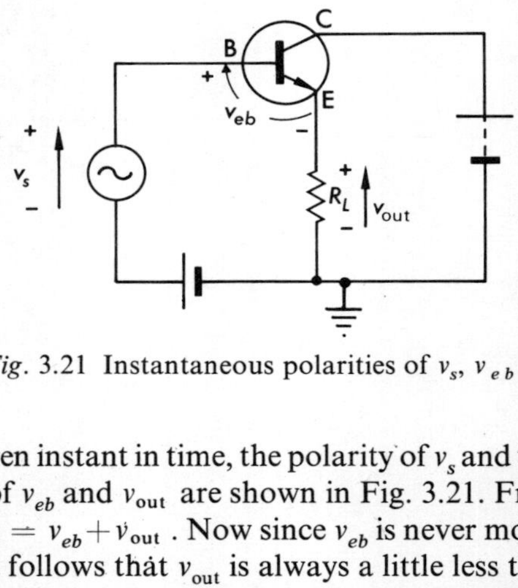

Fig. 3.21 Instantaneous polarities of v_s, v_{eb} and v_{out}.

For a given instant in time, the polarity of v_s and the corresponding polarities of v_{eb} and v_{out} are shown in Fig. 3.21. From this it may be seen that $v_s = v_{eb} + v_{out}$. Now since v_{eb} is never more than a fraction of a volt, it follows that v_{out} is always a little less than v_s. Hence the voltage gain of the emitter follower v_{out}/v_s is always less than unity and a typical value will be between 0·90 and 0·98.

Note: Since $v_s = v_{eb} + v_{out}$, providing v_{eb} is of the order of 0·6–0·7 V, v_s and v_{out} may be quite large. Hence the emitter follower has large signal-handling capacity.

Power gain. By virtue of its high current gain and a voltage gain which is only a little less than unity, the emitter follower has a considerable power gain.

Input and output resistance. Because the whole of the alternating output voltage effectively subtracts from the input signal, it is said that 100 per cent negative feedback occurs. The effect of this negative feedback is to increase the input resistance and decrease the output resistance of the common collector circuit. The actual values of these resistances depend upon:

(a) The signal source resistance, R_s.
(b) The load resistance, R_L.
(c) The forward gain of the transistor, h_{fc}.

For example, $R_{in} \simeq h_{fc} R_L$, and a typical value may lie between 10 kΩ and 500 kΩ, whereas, $R_{out} \simeq R_s/h_{fc}$ and may be as small as 20 Ω. Hence the circuit has a high input resistance and a low output resistance. Note that the values of these resistances are in the reverse order to those of the other two connections.

Hence the special features of the emitter-follower are:

(a) A high current gain.
(b) A voltage gain which is a little less than unity.
(c) Large signal handling capacity.
(d) A significant power gain.
(e) A high input resistance.
(f) A low output resistance.
(g) Output voltage in phase with input voltage.

Because of these properties, the main application of the emitter-follower is as a power amplifying matching device, i.e. it may be connected between a high impedance source and a low impedance load, without excessive loss of power due to a mismatch. (Mismatch merely refers to the condition in which the impedance of a load does not match the impedance of the source to which it is connected. Under such conditions there is less than maximum transfer of energy between source and load).

A summary of the general properties of the three connections is given in Table 3.1.

TABLE 3.1 CB, CE and CC circuit properties

Property	CB	CE	CC or Emitter–follower
Current gain	h_{fb} < unity	h_{fe} (40–200)	h_{fc} (40–200)
Voltage gain	High	High	Less than unity
Power gain	High	High	Moderate
Input resistance	Low (30–50 Ω)	Moderate (500–2 kΩ)	High (10 k–500 kΩ)
Output resistance	High (100–500 kΩ)	Moderate (10–50 kΩ)	Low (20–100 Ω)

3.9 Equivalent Circuit of a Transistor

Like the diode, the transistor may be represented by an equivalent circuit. In fact, since the transistor is more complex than a simple diode, its equivalent circuit is much more useful and more extensively used.

In the past, many different equivalent circuits were used for the transistor and advantages were claimed for all of them. This rather confusing state was due to the seeming complexity of the transistor, which meant that most equivalent circuits were necessary approximations to the true behaviour of the device. It is now general practice, for low-frequency applications, to represent the transistor by the **h-parameter equivalent circuit.** The reason for its general acceptance are:

(a) It is convenient to use in circuit design and analysis.
(b) The parameters used are relatively easy to measure.
(c) The parameters enable the manufacturer to specify the electrical characteristics of the transistor in a convenient way.

This form of equivalent circuit is determined by considering the transistor as contained in a 'black box' which has four terminals, the electrical behaviour of which may be characterized by knowing the relationship between the terminal currents and voltages. Because the characteristic curves actually represent these relationships, the parameters used in the equivalent circuit may be obtained directly from their slopes. Furthermore, by virtue of this 'black box' approach the resulting equivalent circuit applies equally well to all three transistor configurations. In fact this concept may be applied to any four-terminal electrical circuit or system.

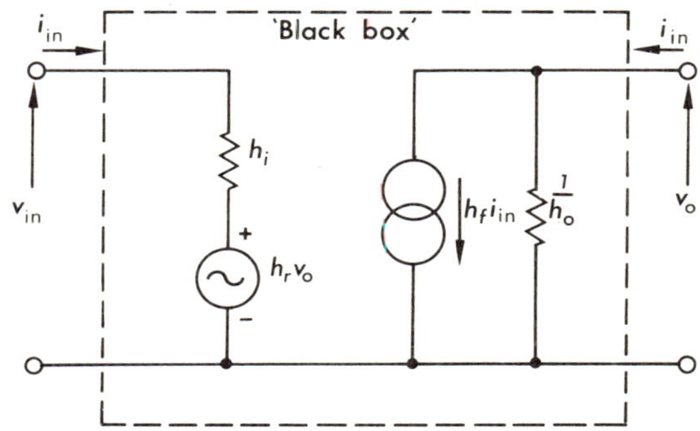

Fig. 3.22 The *h*-parameter equivalent circuit.

Fig. 3.22 shows the general form of the *h*-parameter equivalent circuit. It simply consists of two resistors, an a.c. voltage generator and a current generator. These four component parts represent the following:

(a) The resistor h_i represents the input resistance.
(b) The a.c. voltage generator of e.m.f. $(h_r v_o)$ represents the fraction of the alternating output voltage v_o which is fed back and acts in opposition to the input signal. The parameter h_r is the transistor's feedback factor.
(c) The current generator supplies a current $h_f i_{in}$ to the output circuit. This depicts that the input current i_{in} is amplified and appears in the output circuit as $h_f i_{in}$, where h_f is the forward current gain of the transistor.
(d) The resistor $1/h_o$ represents the output resistance. (Since admittance is the reciprocal of impedance the parameter h_o represents the output admittance.)

The various *h*-parameters are defined as:

$$h_i = \text{input resistance} = \frac{v_{in}}{i_{in}}, \quad \text{with } \mathbf{v}_o = 0.$$

F

It has units of ohms and may be determined from the slope of the input characteristic.

$$h_r = \text{reverse voltage feedback ratio} = \frac{v_{in}}{v_o}, \quad \text{with } i_{in} = 0.$$

It has no units since it is a voltage ratio, the magnitude of which is given by the slope of the voltage feedback characteristic.

$$h_f = \text{forward current gain} = \frac{i_o}{i_{in}}, \quad \text{with } v_o = 0.$$

It has no units since it is a current ratio, the magnitude of which is given by the slope of the transfer characteristic.

$$h_o = \text{output admittance} = \frac{i_o}{v_o}, \quad \text{with } i_{in} = 0.$$

It has the units of siemens and maybe determined from the slope of the output characteristic.

A second subscript *e*, *b*, or *c* is added to indicate that the transistor is connected in CE, CB, or CC configuration. For example, $h_{ie} = 800\ \Omega$ would indicate that the input resistance is 800 Ω when connected in the CE mode, or $h_{ib} = 20\ \Omega$ would indicate that the input resistance is 20 Ω when connected in the CB mode.

The name hybrid is used for the parameters because of the mixture or hybrid nature of their units, i.e. h_i—ohms, h_r—voltage ratio, h_f—current ratio and h_o—siemens.

Fig. 3.23 shows a transistor connected in the CE and CB modes, enclosed in a 'black box' and their equivalent hybrid parameter circuits. Obviously the CC mode could have been equally well represented. Typical numerical values for a low power general purpose transistor are: $h_{ib} = 30\ \Omega$, $h_{rb} = 4.10^{-4}$, $h_{fb} = -0{\cdot}98$ and $h_{ob} = 10^{-6}$ S, for the CB mode. It must be remembered that all the values quoted above vary considerably between transistors of the same type number and with the condition under which they are measured. The corresponding numerical values for the CE and CC connection may be calculated from Table 3.2 (where $\Delta_b = h_{ib}h_{ob} - h_{rb}h_{fb}$).

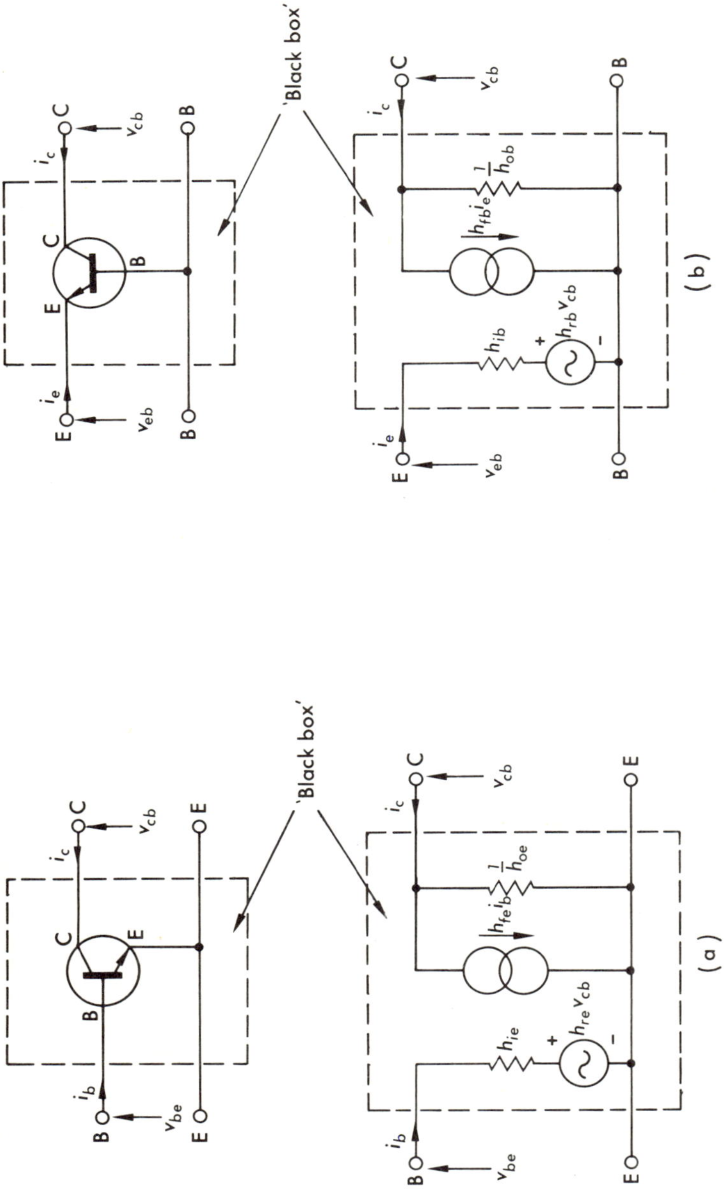

Fig. 3.23 Hybrid equivalent circuits for CE and CB configuration.
(a) CE connection.
(b) CB connection.

TABLE 3.2 Relationship between h-parameters

CB mode	CE mode	CC mode
h_{ib}	$h_{ie} = \dfrac{h_{ib}}{1 + h_{fb}}$	$h_{ic} = \dfrac{h_{ib}}{1 + h_{fb}}$
h_{rb}	$h_{re} = \dfrac{\Delta_b - h_{rb}}{1 + h_{fb}}$	$h_{rc} = 1$
	$h_{fe} = \dfrac{-h_{fb}}{1 + h_{fb}}$	$h_{fc} = \dfrac{-1}{1 + h_{fb}}$
h_{ob}	h_{oe} $\dfrac{h_{ob}}{1 + h_{fb}}$	h_{oc} $\dfrac{h_{ob}}{1 + h_{fb}}$

3.10 Electrical Characteristics and Maximum Ratings

By the use of differing fabrication techniques and by variation in physical dimensions, transistors are constructed for widely differing applications. Manufacturers supply with their products data sheets which give a comprehensive review of all the information that is necessary to use their devices. Naturally for some applications some of this information is more relevant than the rest. In general, however, the engineer will always need to know a few basic parameters and ratings concerning the transistor in order to operate it correctly. Such characteristics and ratings are given in Tables 3.3 and

TABLE 3.3 Major characteristics for a small signal transistor

Parameter	Symbol	Min.	Typical	Max.	Unit	Test conditions
Small-signal h-parameters	h_{ie}	—	500	1000	Ω	$V_{CB} = 5$ V
	h_{re}	—	$0{\cdot}85\ 10^{-4}$	$5{\cdot}0 10^{-4}$	—	$I_C = 1$ mA
	h_{oe}	—	35	80	μS	
	h_{fe}	20	100	—	—	At 1 kHz
Gain–band-width product	f_T	—	30	50	MHz	

3.4. The values quoted may be regarded as being typical for a small-signal, general-purpose, npn planar silicon transistor at an ambient temperature of 25°C.

Note: The gain–bandwidth product f_T defines the upper frequency limit of the transistor, i.e. f_T is the frequency at which the common-emitter current gain has fallen to unity.

TABLE 3.4 Major maximum ratings for a small signal transistor

Maximum rating	Symbol	Value	Unit
Collector-to-base voltage	V_{CB}	60	V
Collector-to-emitter voltage	V_{CE}	45	V
Emitter-to-base voltage	V_{EB}	6	V
Collector current	I_C	30	mA
Collector power dissipation	P_{tot}	300	mW

3.11 Application of a Transistor as an Amplifier

From the previous considerations of the three possible connections of a transistor it is apparent that for both current and voltage amplification the common-emitter connection is by far the most useful.

In order for the transistor to operate correctly and efficiently as an amplifier, suitable d.c. operating conditions must be established. These d.c. conditions involve more than merely ensuring that the emitter–base junction is forward biased and the collector–base junction is reverse biased. They must also ensure:

(a) that the transistor operates over the linear part of its characteristic—if this is not achieved, then the output signal will be distorted, i.e. it will not be a true magnified replica of the input signal;

(b) that the no-signal operating point remains fixed in spite of variations in collector current caused by variation of leakage current with temperature;

(c) that the manufacturer's maximum ratings for the transistor are not exceeded when an alternating input signal is superimposed on the existing d.c. conditions.

Before considering an actual amplifier the above points will be briefly examined.

3.11.1 *The d.c. load line and operating point*

Fig. 3.24 shows a simple CE amplifier, together with the output characteristic of the transistor. Consider variations of the collector–emitter voltage V_{CE} with variation of collector current I_C due to the base current I_B changing. When I_B is zero, I_C will also be zero, since $I_C = h_{fe} I_B$, (neglecting the small leakage current I_{CEO}). Therefore the voltage drop across the collector load resistor $I_C R_L$ will also be zero. This means that the whole of the 12 V of the collector supply voltage is dropped across the transistor. Thus when $I_C = 0$, $V_{CE} = 12$ V—this is represented on the output characteristic by the point M. If I_B increases I_C follows, causing corresponding increases in the voltage dropped across R_L and corresponding decreases in the voltage across the transistor. Eventually the whole of the 12 V will appear across R_L (neglecting the small saturation resistance of the transistor). In this case

$$I_C = \frac{12}{4.10^3} = 3 \text{ mA}, \qquad V_{CE} = 0.$$

This is represented on the output characteristic by the point N.

Although only the two extreme conditions have been considered, i.e. when $I_C = 0$ and when $V_{CE} = 0$; by joining the two points M and N so obtained, the resulting straight line, called the **d.c. load line** represents all possible pairs of values of I_C and V_{CE} for this particular amplifier. For example, if $I_B = 30\ \mu$A then by projecting vertically and horizontally from the point where the load line intersects the characteristic for $I_B = 30\mu$A, i.e. point P, to the two axes, the corresponding values of V_{CE} and I_C are seen to be 5 V and 1·75 mA respectively.

The importance of the d.c. load line is not limited to the above. It also indicates the best possible d.c. operating point for the amplifier,

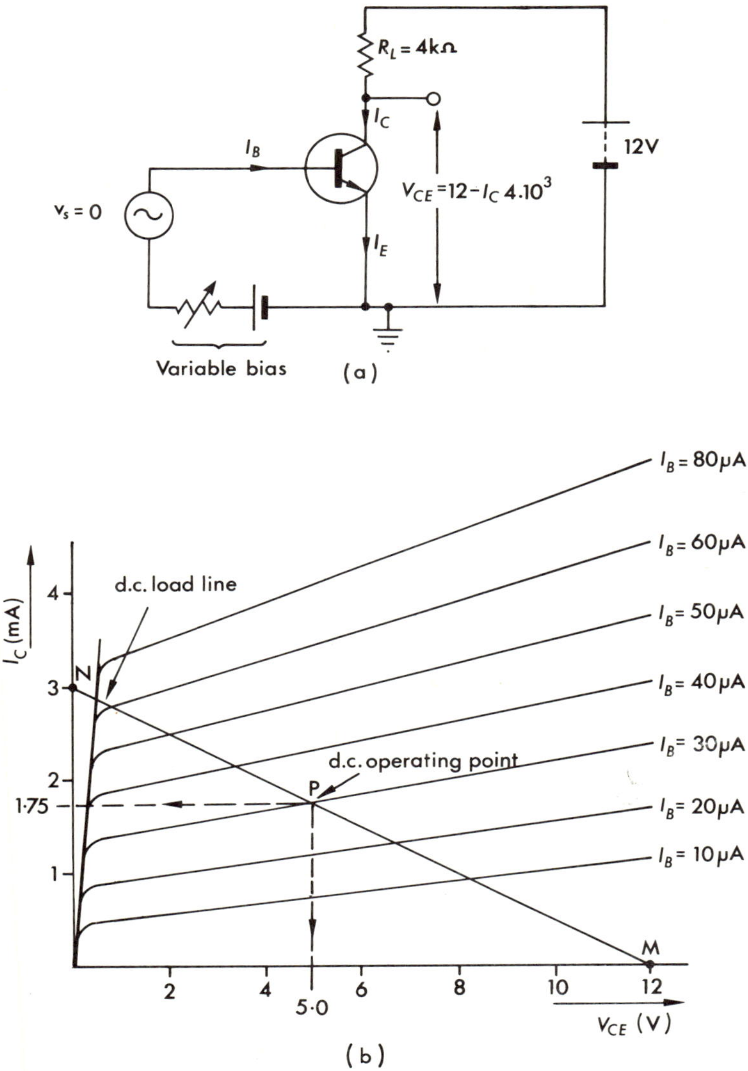

Fig. 3.24 The d.c. load line for an amplifier.
(a) Simple amplifier.
(b) Output characteristic and d.c. load line.

i.e. the optimum bias condition for the transistor under no-signal conditions. Furthermore, on inspection of the output characteristic and the load line, the maximum amplitude of input signal which the amplifier can accommodate may be seen.

For most amplifying applications the d.c. operating point is taken approximately midway along the load line (point P). This simply means in this particular case that in the absence of an input signal, the d.c. base current is arranged to be 30 μA resulting in I_C and V_{CE} being 1·75 mA and 5V respectively. These are the d.c. quiescent (no-signal) conditions and point P is known as the **d.c. operating** or **quiescent point** of the amplifier. Fig. 3.24(b) shows that the maximum amplitude of input signal that this amplifier can handle without distortion (see later in this section) is that which causes the base current to vary from 0 to 60 μA.

When inserting the d.c. load line on the output characteristic in the last section, it was assumed for simplicity that both the leakage current and the saturation resistance of the transistor were negligible. Although both are small, in practice they are taken into account otherwise distortion of the signal may result.

In Fig. 3.25, three shaded regions are indicated; they are:

(a) **Saturation Region.** This region occurs between the saturation line and the vertical axis, i.e. for low values of V_{CE}. When a transistor saturates, I_C no longer varies in sympathy with I_B; thus the output signal will not be a true magnified version of the input signal. In other words, any fraction of the input signal which causes the dynamic or instantaneous operating point to enter this saturated region will be clipped or lost to the output.

(b) **Cut-off Region.** This region occurs between $I_B = 0$ and the horizontal axis. In order to enter this region the emitter–base junction must become reverse biased and in consequence, as the name suggests, the transistor is cut off. Any fraction of the input signal which causes this to occur will be lost to the output.

(c) **Maximum power dissipation curve.** The heat produced within a transistor is almost entirely generated at the collector junction. Power dissipation at the emitter junction is $I_E \times V_{BE}$ and is $I_C \times V_{CE}$ at the collector junction. It follows that since

$I_C \simeq I_E$ and $V_{CE} \gg V_{BE}$ then the total power dissipation P_{tot} of the transistor is rated as $(I_C \times V_{CE})$, the small power dissipated at the emitter junction being ignored.

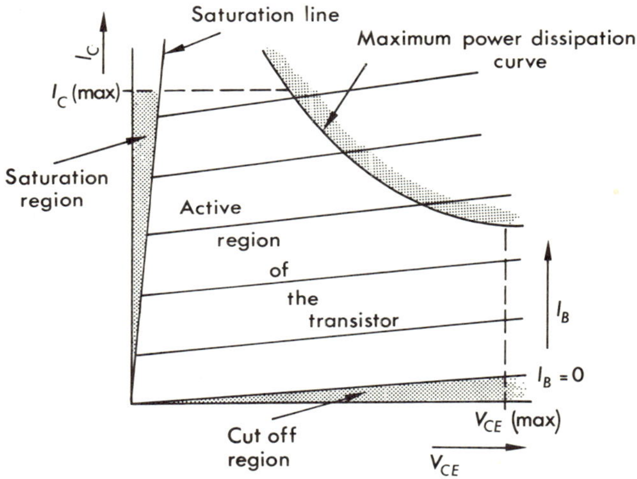

Fig. 3.25 Operating region of the transistor.

The heat developed at the collector junction raises the junction temperature. If this temperature is permitted to increase, I_{CEO} will increase so causing I_C to follow, which will cause further heat to be dissipated at the junction. The action is cumulative resulting in **thermal runaway** and eventual destruction of the transistor. In order to avoid this, manufacturers always quote P_{tot} in terms of ambient and junction temperature (symbols T_{amb} and T_{γ} respectively) for every type of transistor. Thus if P_{tot} is known and $P_{tot} = I_C V_{CE}$, then by assuming values for V_{CE} the corresponding maximum value of I_C may be calculated. If these pairs of values are plotted on the output characteristic, the result is the maximum power dissipation curve for the transistor. It is essential, for the above reasons, that the transistor is never operated beyond the limit imposed by this curve.

Finally, for reasons given earlier in the chapter, I_C(max) and V_{CE}(max) must not be exceeded.

3.11.2 *d.c. bias stabilization*

There are many different ways in which the correct d.c. bias may be obtained but the most widely used is the circuit shown in Fig. 3.26.

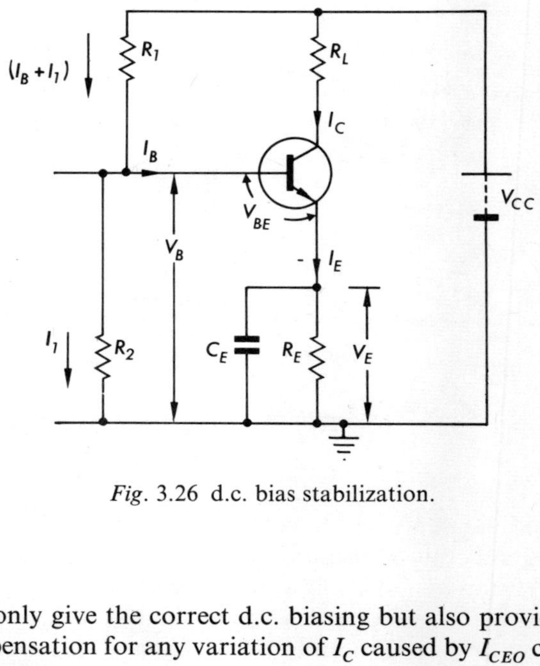

Fig. 3.26 d.c. bias stabilization.

It will not only give the correct d.c. biasing but also provide automatic compensation for any variation of I_C caused by I_{CEO} changing with temperature.

The base is connected to the junction of R_1 and R_2 which form a potential divider across the supply voltage. By making the bleed current I_1 much larger than the d.c. base current $I_B (I_1 \geqslant 10_1 I_B)$, the voltage at the base V_B is dependent on R_1 and R_2 and practically independent of I_B. The emitter voltage V_E is dependent on I_E and R_E. By careful choice of R_1, R_2 and R_E, the base–emitter junction will receive the correct forward bias.

Stabilization against temperature change is obtained as follows:

if I_C increases due to I_{CEO} increasing with temperature, then I_E must also increase, causing V_E to increase. Since V_B is practically independent of transistor currents, it will remain constant. Hence if V_E increases and V_B is constant, then their difference V_{BE} must decrease, i.e. $V_{BE} = V_B - V_E$. Any decrease of the forward bias voltage V_{BE} must result in a corresponding decrease of I_C. Thus the original increase in I_C due to I_{CEO} is offset.

3.11.3 *Current, voltage and power gain*

Consider the application of a sinusoidal input signal to the base of the amplifier shown in Fig. 3.24. This alternating signal will be superimposed on the existing d.c. conditions. Let its magnitude be ± 50 mV peak which causes the base current to vary, say, ± 20 μA.

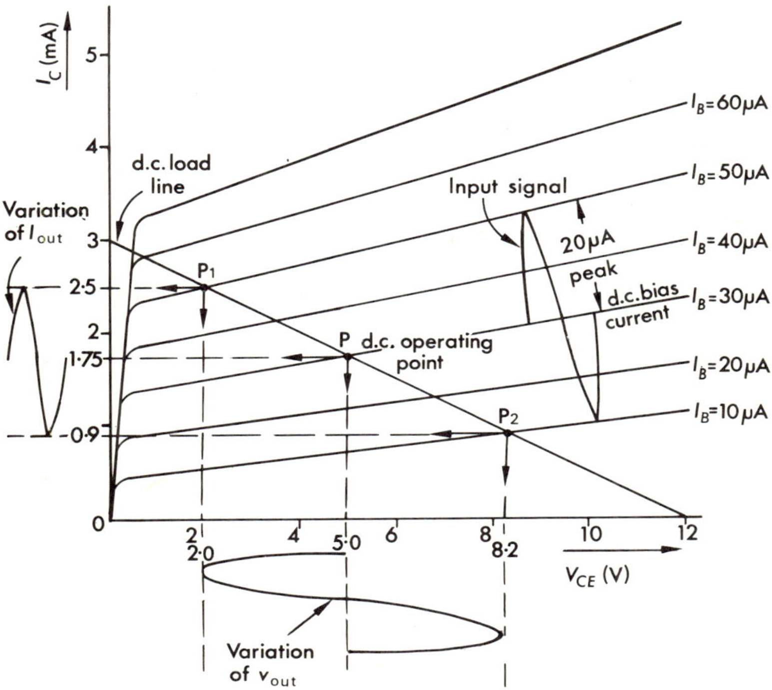

Fig. 3.27 Variation of output current and voltage for a given input signal.

The output characteristic of the transistor and the d.c. load line are redrawn in Fig. 3.27, together with the input signal and the resulting collector current and voltage waveforms.

In the absence of a signal, the base-bias current is 30μA and the d.c. operating point corresponds to $I_C = 1{\cdot}75$ mA, $V_{CE} = 5{\cdot}0$ V. When the input signal is on a positive peak the base current increases to a maximum value of 50 μA (= d.c. bias + a.c. signal). The dynamic operating point of the amplifier moves up the load line to P_1. By projecting horizontally and vertically from P_1 to the two axes, the corresponding values of I_C and V_{CE} are 2·5 mA and 2·0 V respectively. Conversely, when the input signal is on a negative peak, the base current decreases to a minimum of 10 μA (=d.c.bias − a.c. signal). The dynamic operating point of the amplifier moves down the load line to P_2. Again by projection the corresponding values of I_C and V_{CE} are 0·9 mA and 8·2 V respectively.

Although only three conditions of the input and the resulting output have been considered, it is obvious that for every instantaneous value of the base current, the corresponding values of I_C and V_{CE} may be found by projection. The resulting waveforms for both I_C and V_{CE} for a given input waveform are illustrated in Fig. 3.27.

Current gain (A_i). From Fig. 3.27 it may be seen that the alternating base current i_{in} varies from 10 to 50 μA. The resulting variation in the collector current i_{out} is (2.5–0·9) = 1·6 mA peak-to-peak.

$$\text{Current gain, } A_i = \frac{i_{out}}{i_{in}} = \frac{1{\cdot}6 \,.\, 10^{-3}}{40 \,.\, 10^{-6}} = 40 \text{ times.}$$

Voltage gain (A_v). An alternating input voltage v_{in} of ±50 mV peak produces an alternating output voltage v_{out} of (8·2–2·0) = 6·2 V peak-to-peak.

$$\text{Voltage gain, } A_v = \frac{v_{out}}{v_{in}} = \frac{6{\cdot}2}{100 \,.\, 10^{-3}} = 62 \text{ times.}$$

Power gain (A_p). Power gain is the product of current and voltage gains.

$$\text{Power gain, } A_p = A_i \times \mathbf{A}_v$$
$$= 40 \times 62$$
$$= 1480 \text{ times.}$$

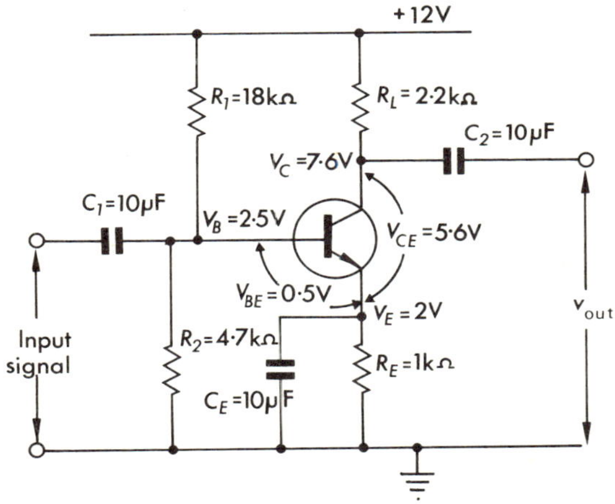

Fig. 3.28 AF amplifier.

3.11.4 *Practical audio-frequency amplifier*

Fig. 3.28 shows a typical low-power audio-frequency (AF) amplifier. The various components have been given numerical values merely as an indication of typical magnitudes. They will of course vary considerably, depending upon the type of transistor and required d.c. conditions.

The functions of the various components are as follows: C_1 couples the alternating input signal to the base of the transistor, while at the same time acting as a d.c. block, i.e. preventing any d.c. component of the signal from upsetting the biasing of the transistor. Resistors R_1, R_2 and R_E set up the required biasing, as well as providing stabilization against I_{CEO}. R_L is, of course, the transistor load resistor. C_E bypasses the emitter resistor R_E. This is necessary in order to avoid degeneration caused by negative feedback, which would otherwise be developed across R_E (see section 3.8). C_2 acts in a similar way to C_1, except that it is the amplified output signal which is passed.

REFERENCES

ABRAHAMS, J. R. and G. J. PRIDHAM. *Semiconductor Circuits: (Theory Design and Experiment)* Pergamon Press Ltd., 1966.

BARDEEN, J. and W. H. BRATTAIN. 'The Transistor, a Semiconductor Triode', *Phys. Rev.,* **74**, 230, 1948.

DEAN, K. J. *Transistors Theory and Circuitry*, McGraw-Hill Book Company, 1964.

GIBBONS, J. F. *Semiconductor Electronics*, McGraw-Hill Book Company, 1966.

MANN, G. B. *ABC's of Transistors*, W. Foulsham & Co. Ltd., 1968.

PIKE, C. A. *Basic Transistor Circuits*, W. Foulsham & Co. Ltd., 1969.

Chapter 4

Junction Field-Effect Transistors

4.1 Introduction

Field-effect transistors (usually abbreviated to FET) may be divided into two main categories:

(a) **Junction FETs.**
(b) **Insulated gate FETs.**

The former will be considered in this chapter and the latter in the next.

Basically a junction FET is a slice of silicon whose conductance is controlled by an electric field acting perpendicularly to the current path. This electric field results from a reverse-biased pn junction and because of the importance of this transverse field the device is so named.

The junction FET differs from the conventional transistor in a number of distinct ways:

(a) Its principle of operation.
(b) Current is carried by only one type of carrier—majority carriers. This is different from conventional transistors where both majority and minority carriers are involved. Because of this, the FET is often referred to as a **unipolar transistor** and the conventional type as a **bipolar transistor.**
(c) The junction FET exhibits a high input impedance. This is a distinct advantage over the relatively low input impedance of the bipolar transistor.
(d) Because of their high input impedance, junction FETS are voltage operated as opposed to the current-operated bipolar transistor.

Further differences, including advantages and disadvantages will be discussed later in this chapter.

Figure 4.1 shows the schematic representation and graphical symbol of the junction FET. Although the schematic diagram does not resemble the physical structure of the junction FET, it is admirably suited for examining the principle of its operation. Notice that the junction FET is like the bipolar transistor, in that p-type and n-type regions may be interchanged.

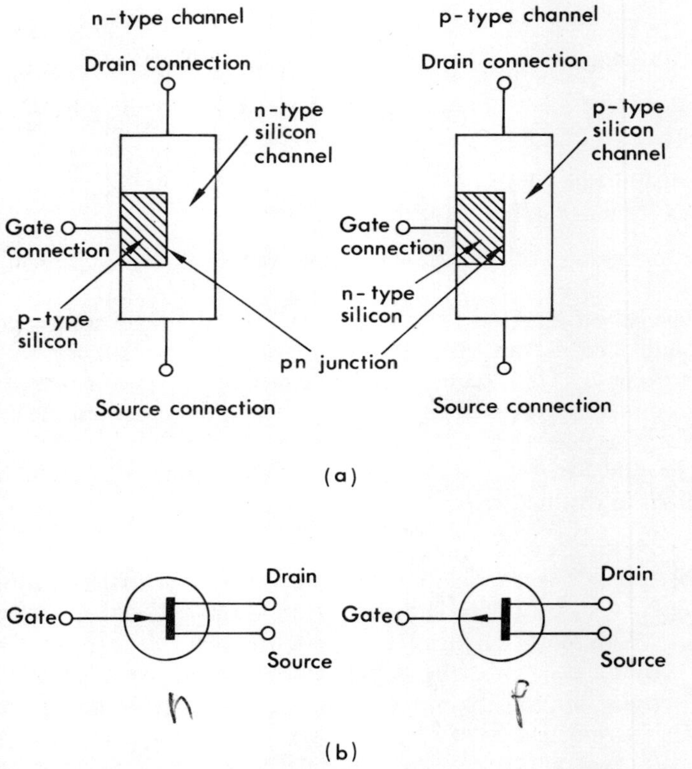

Fig. 4.1 (a) Schematic representation of a junction FET.
(b) The graphical symbols for n- and p-type channel FETS.

The direction of the arrow at the gate of the graphical symbol of the FET indicates the direction in which the gate current would flow if the gate junction were forward biased.

4.2 Basic Structure of the Junction FET

In order to appreciate the operation of the junction FET, it is helpful to examine one form of its basic construction, as shown in Fig. 4.2.

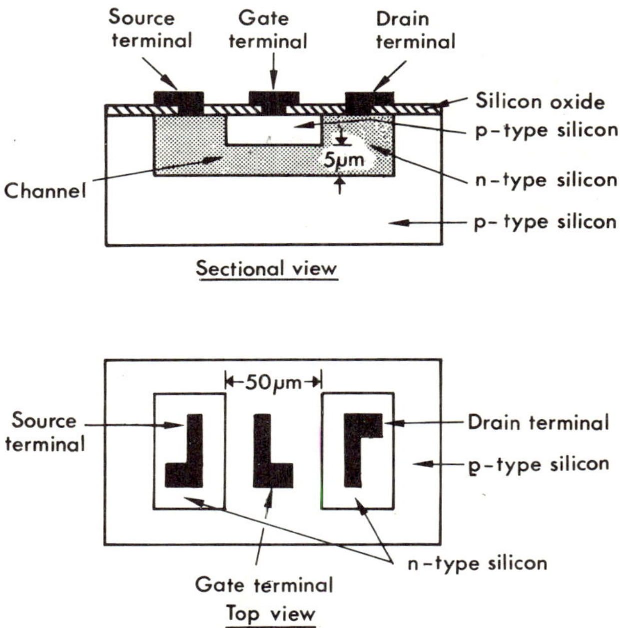

Fig. 4.2 Structure of a junction FET.

The n-channel junction FET consists essentially of an n-type region, called the **channel,** which is formed into a p-type **substrate** or wafer of silicon. A second p-type region is formed in the n-type channel so that the resulting U-shaped channel is sandwiched between two p-type regions. The two p-type regions are made common at the end of the structure. Contacts are made to the two ends of the channel, one of which is known as the **source** and the other the **drain.** A contact to the p-type region called the **gate** completes the three external connections to the device.

G

In order to convey some idea of the extremely small size of the structure, approximate dimensions of the active part of the channel have been included in Fig. 4.2.

4.3 Operation of a Junction FET

As explained in section 2.2, at the common boundary between p-type and n-type semiconductor materials there is always a very thin depletion layer separating them. Furthermore, the thickness of this depletion layer varies in accordance with the magnitude and polarity of any voltage which may be applied across the junction. The operation of the junction FET depends very largely upon this variation and so the behaviour of the depletion layer under the influence of different terminal potentials will be examined in detail.

The potentials applied to the terminals of a FET (common-source mode) are referred to the source. For example, $V_{GS} = -2$ V, indicates that the gate potential is 2 V negative with respect to the source, or $V_{DS} = 10$ V, indicates that the drain potential is 10 V positive with respect to the source.

4.3.1 *Absence of bias voltages*

Consider an n-type channel FET with no voltages applied to any of

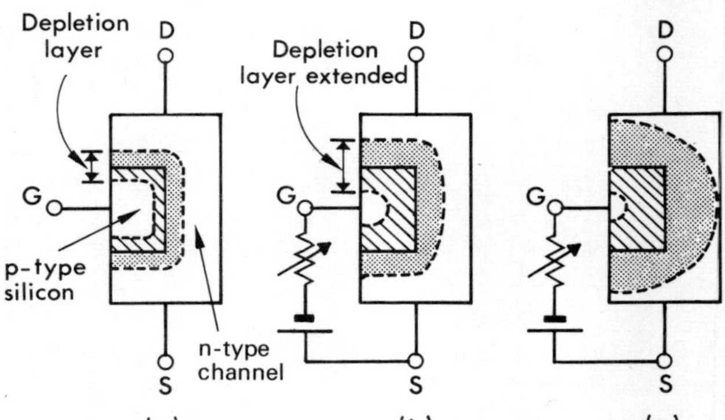

Fig. 4.3 Effects of V_{GS} on the depletion layer.
(a) $V_{GS} = 0$, $V_{DS} = 0$ (b) $V_{GS} > 0$, $V_{DS} = 0$ (c) $V_{GS} = V_C$, $V_{DS} = 0$

the three connections. The depletion layer associated with the pn junction is very thin and evenly distributed around it (Fig. 4.3a).

4.3.2 *Gate potential but no V_{DS}*

With $V_{DS} = 0$ and the pn junction reverse-biased (negative potential to p-type and positive potential to n-type), the depletion layer width is extended, as explained in section 2.2 (see Fig. 4.3b). As

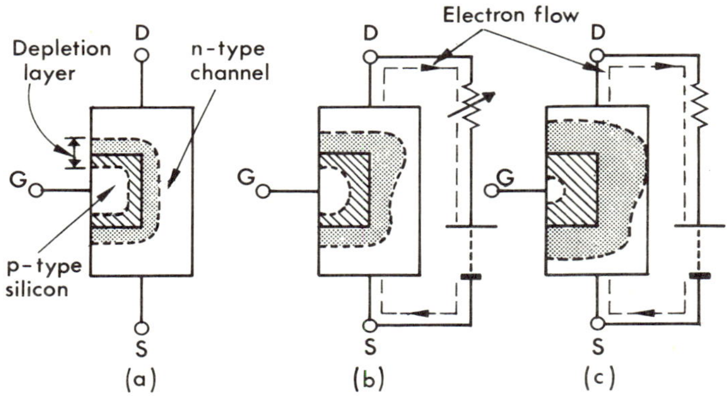

Fig. 4.4 Effects of V_{DS} on the depletion layer.
(a) $V_{GS} = 0$, $V_{DS} = 0$ (b) $V_{GS} = 0$, $V_{DS} > 0$ (c) $V_{GS} = 0$, $V_{DS} = V_P$

V_{GS} is increased, the width of the depletion layer extends into the channel until eventually, providing the initial channel width is sufficiently small, the depletion layer extends fully across the channel region (Fig. 4.3c). The channel is said to be 'cut-off' and the required value of V_{GS} to achieve this is V_c, the cut-off potential.

4.3.3 *Drain potential but no V_{GS}*

With $V_{GS} = 0$ and the drain positive with respect to the source the majority carriers, i.e. electrons within the n-type channel, will be repelled by the negative potential at the source and attracted by the positive potential at the drain. Therefore electrons will flow from the source to the drain and a corresponding current will flow in the external circuit.

The actual magnitude of this current will depend upon four factors:

(a) The number of electron carriers available within the channel, i.e. the conductivity of the channel. This will of course be controlled by the level of doping of this region.
(b) the length l of the channel.
(c) The cross sectional area A of the conducting portion of the channel.
(d) The magnitude of the applied voltage V_{DS}.

Thus the junction FET acts as a silicon resistor whose resistance R is given by $\rho l/A$ and the magnitude of the current (I_D) for low values of V_{DS} is given by V_{DS}/R ($V_{GS} = 0$).

Owing to the resistance of the channel and the applied voltage V_{DS}, a potential gradient will exist along the channel (Fig. 4.5).

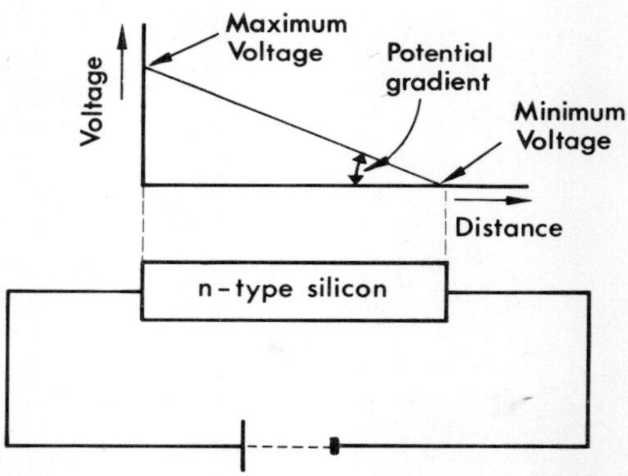

Fig. 4.5 Potential gradient along a piece of n-type silicon.

This means that the pn junction becomes progressively more reverse biased as the drain is approached, i.e. the channel potential with respect to the gate increases as the drain is approached. Consequently the depletion layer takes on a wedge shape, such as that depicted in Fig. 4.4(b).

As V_{DS} increases, the wedge-shaped depletion layer increases until at a certain value of V_{DS} the channel is *almost* completely pinched off. This value of V_{DS} is called the **pinch-off voltage** V_P.

However, the pinch-off voltage does not completely cut off the drain current I_D; this is because of the electrostatic field associated with the electrons making up I_D. This electric field produces repulsion forces and there is always a tendency for the electrons to spread out; but owing to the widening of the depletion layer they are forced to flow within a confined space which becomes progressively smaller. Eventually the constriction imposed on the electrons reaches a limit. Beyond this, their concentrated electric field successfully opposes further extension of the depletion layer. Hence I_D cannot be completely cut off. Note the difference between V_P and V_C.

4.3.4 *Effect of both V_{GS} and V_{DS}*

Having examined in some detail the behaviour of the depletion layer under the influence of V_{GS} and V_{DS} independently, the actual operation of the junction FET under their simultaneous influence will be now considered using a typical output characteristic for the device.

4.4 Output Characteristic

As shown by Fig. 4.6(b), with $\mathbf{V}_{GS} = 0$ the current through the FET increases almost linearly until V_{DS} approaches V_P. This indicates that the device is acting as a simple voltage controlled resistor, the value of the resistance being given by the reciprocal of the slope. Further increase in V_{DS} greater than V_P results in only a slight increase in I_D. Thus the drain current has reached its saturation value because the channel has become pinched off and V_{DS} has lost control of I_D.

Although the FET is said to be saturated when V_{DS} is greater than V_P, it must be remembered that this phenomenon is different from that of the bipolar transistor. For example, the saturation state of a bipolar transistor is normally avoided, since in this state a change in input current does not produce a corresponding change in output current. Furthermore saturation occurs for low

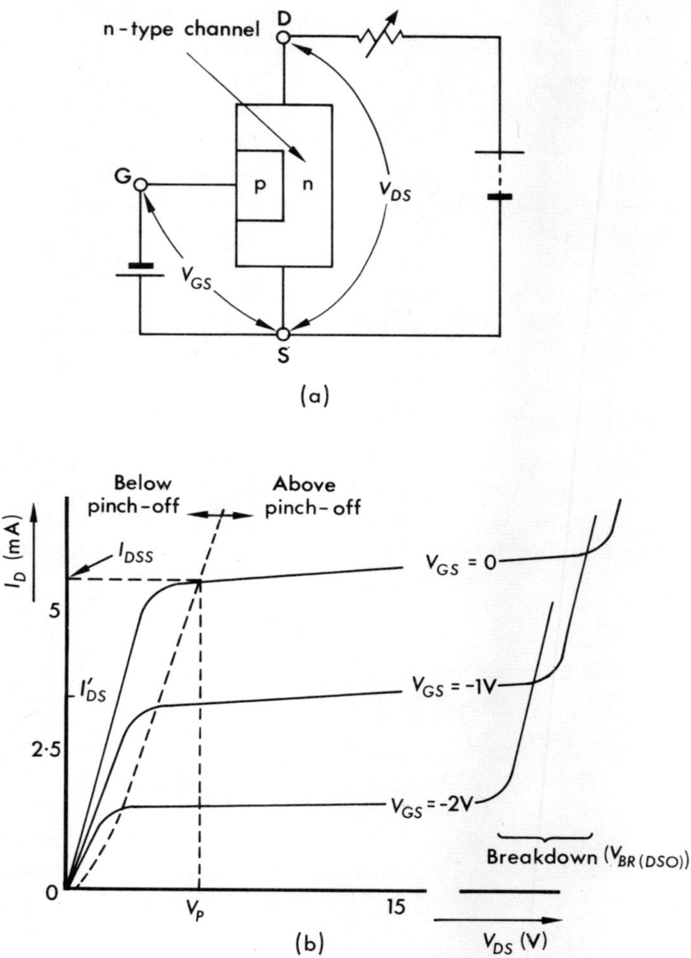

Fig. 4.6 Biasing voltages and output characteristic of a junction FET.
(a) Biasing.
(b) Output characteristic.

values of output voltage and the maximum value of the output current is controlled only by the resistance in the external circuit. In contrast to this, the FET normally operates in the saturated region of its characteristic, i.e. $V_{DS} > V_P$, where the output current is controlled by the input potential. Also the pinch-off effect and not the resistance of the external circuit exerts the main control over the maximum output current. Hence the saturation phenomena associated with bipolar and unipolar devices have little in common.

Over the saturation region, the FET exhibits a much larger resistance, this is the **drain** (or **output**) **resistance** r_d which is given by the reciprocal of the slope of the output characteristic and is defined by

$$r_d = \frac{\Delta V_{DS}}{\Delta I_D} \qquad (V_{GS} \text{ constant})$$

Typically r_d is of the order 100 kΩ to 1 MΩ.

Note from Fig. 4.6(b) that the slope of the characteristic for $V_{DS} > V_P$ progressively increases as V_{GS} increases, indicating that the drain resistance is decreasing. The slope of the characteristic is the drain conductance g_d and is sometimes referred to as the output conductance g_{DS} or output admittance y_{DS}.

If the gate potential is increased to $V_{GS} = -1V$, the depletion layer width will be increased thereby reducing the cross sectional area of the conducting channel. Thus when V_{DS} is applied, the resulting current I_D which will flow will be less than the corresponding I_D for $V_{GS} = 0$, i.e. the effect of V_{GS} and V_{DS} on extending the depletion layer into the channel is additive. The increased resistance of the channel may be found by determining the slope of characteristic for $V_{GS} = -\mathbf{1}$ V and $V_{DS} < V_P$. Furthermore, due to the simultaneous additive effect of both $V_{GS} = -1$ V and $V_{DS} = V_P$ on the depletion layer, the drain current saturates at a lower value, I'_{DS}, compared to I_{DSS}.

By varying V_{GS} and plotting I_D against V_{DS}, a family of output characteristics may be obtained. Note in Fig. 4.6(b) that the characteristics for various values of V_{GS} do not merge into a single line as is the case with the bipolar transistor.

A further property which may be seen from Fig. 4.6(b) is that the breakdown voltage $V_{BR(DSO)}$ decreases as V_{GS} becomes more negative. This is caused by the fact that the reverse biased gate potential adds to the drain potential, so that the effective voltage across the gate junction increases and at $V_{BR(DSO)}$ the junction breaks down and the current avalanches. (The triple subscript is used, so that $V_{BR(DSO)}$ represents the **br**eakdown voltage of the **d**rain–gate junction with the **s**ource **o**pen.

4.5 Mutual or Transfer Characteristic

When a FET is used in the common-source mode, the gate is the control electrode, rather like the base in a bipolar transistor when connected in the CE mode. The output characteristic shows that I_D is controlled by V_{GS}, for V_{DS} greater than V_P. Thus when a signal is applied to a FET, it is introduced across the gate and the source. Now since the pn junction between the gate and the channel is reverse biased, it follows that the input resistance of the device will be very high, i.e. the resistance 'seen by the signal' will be that of a reverse biased pn junction, which may be as high as 1000 MΩ at zero frequency.

In many respects a FET is similar to a pentode valve. Both have input resistances and high amplification and both are voltage controlled. Just as varying the voltage on the control grid of a pentode will control the anode current, so the voltage on the gate controls the drain current of a FET.

If V_{DS} is maintained constant at some value greater than V_P, and a graph of I_D against V_{GS} is obtained, then the resulting curve is called the mutual or transfer characteristic. Fig. 4.7 illustrates such a curve. For interest, a comparison between a FET and a pentode is made.

In a pentode the parameter 'mutual conductance' indicates the control that the grid voltage has over the anode current, and in the FET the *mutual conductance* (g_m) indicates the gate's control over I_D. The actual value of g_m is given by the slope of the transfer characteristic, and is defined by

$$g_m = \frac{\Delta I_D}{\Delta V_{GS}} \qquad (V_{DS} \text{ constant}).$$

The characteristic indicates that the g_m varies with V_{GS}. The actual value of g_m for a particular bias voltage V_{GS} may be determined as follows. The shape of the transfer characteristic is very nearly a parabola and it may be represented by the expression

$$I_D = I_{DSS}\left[1 - \frac{V_{GS}}{V_P}\right]^2 \tag{4.1}$$

where I_{DSS} is the drain current at zero gate-to-source voltage, V_{GS} is the particular value of gate bias for which the corresponding value of g_m is desired.
and V_P is the pinch-off voltage.

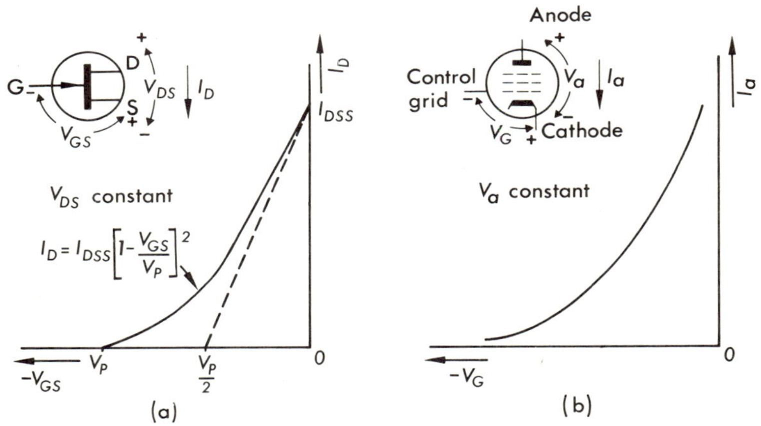

Fig. 4.7 Mutual or transfer characteristic.
(a) Junction FET.
(b) Pentode valve.

By definition g_m is the change of I_D with V_{GS}. Therefore by differentiating equation (4.1) it may be shown that

$$g_m = \frac{2I_{DSS}}{V_P}\left[1 - \frac{V_{GS}}{V_P}\right]. \tag{4.2}$$

Hence if V_P and I_{DSS} are known, g_m may be calculated. Note from equation (4.1) or alternatively from Fig. 4.7(a) that the g_m is a maximum at $V_{GS} = 0$ and zero at $V_{GS} = V_P$. The variations between these points is expressed by equation (4.2). In practice g_m varies considerably for different types of junction FET; a typical range of value for g_m at $V_{GS} = 0$ is from 0·5 to 6 mA/V, although some general-purpose devices have g_m as high as 20 mA/V.

Since g_m has the units of an admittance (A/V) it is also referred to as the **forward transfer admittance** or **forward transadmittance** and designated by the symbols y_{fs} or g_{fs}. In this case the above range of typical values may be expressed as 500 to **6000** μS.

4.6 Parameters g_m, r_d and μ

With a junction FET there are two variable voltages V_{DS}, V_{GS} and a variable current I_D. Any one of these may be fixed and the relationship between the other two found. These relationships may be determined by the use of the parameters g_m, r_d and μ.

Although the mutual conductance g_m and drain resistance r_d have already been defined, for convenience they will be repeated and the third parameter, called the **amplification factor** μ is defined as

$$\mu = \frac{\Delta V_{DS}}{\Delta V_{GS}} \qquad (I_D \text{ constant}) \qquad (4.3)$$

$$g_m = \frac{\Delta I_D}{V_{GS}} \qquad (V_D \text{ constant}) \qquad (4.4)$$

$$r_d = \frac{\Delta V_{DS}}{\Delta I_D} \qquad (V_{GS} \text{ constant}) \qquad (4.5)$$

From the transfer characteristic it may be seen that a change of V_{GS} causes a corresponding change in I_D, for a constant value of V_{DS}. Thus from equation (4.4), $\Delta I_D = g_m \Delta V_{GS}$. From the normal operating region of the output characteristic it may also be seen that I_D decreases slightly with decrease in V_{DS} for constant value of V_{GS}. Thus from equation (4.5),

$$\Delta I_D = \frac{1}{r_d} \Delta V_{DS}.$$

In an amplifier, any variation of V_{GS} causes I_D to change which in turn causes V_{DS} to vary, e.g. in an n-channel junction FET, if V_{GS} increases in a positive direction, I_D increases, which increases the voltage drop across the drain load resistor, resulting in a decrease of drain voltage. The two effects act simultaneously so the total change of I_D is given by

$$\Delta I_D = g_m \Delta V_{GS} - \frac{1}{r_d} \Delta V_{DS} \tag{4.6}$$

This equation is referred to as the **dynamic equation of the junction FET,** since it describes the behaviour of ΔI_D in terms of the devices' parameters and terminal voltage changes ΔV_{GS} and ΔV_{DS}.

If it is arranged that an increase in V_{GS} is offset by a decrease in V_{DS}, then the resulting change in I_D will be zero, so that

$$0 = g_m \Delta V_{GS} - \frac{1}{r_d} \Delta V_{DS}$$

which gives, for a fixed value of I_D,

$$\frac{\Delta V_{DS}}{\Delta V_{GS}} = g_m r_d.$$

However, since by definition $\Delta V_{DS}/\Delta V_{GS} = \mu$ with I_D constant, it follows that

$$\mu = g_m r_d. \tag{4.7}$$

Figs. 4.6 and 4.7 indicate that both r_d and g_m vary with I_D. From this, it is reasonable to expect μ to vary with I_D. Typical variation of the three parameters with I_D is illustrated in Fig. 4.8.

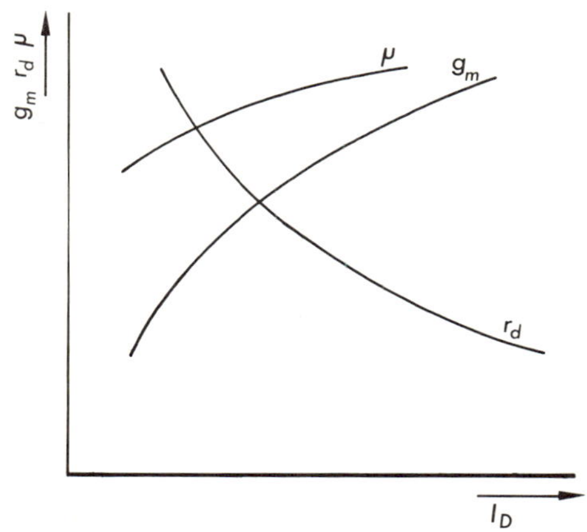

Fig. 4.8 Variation of g_m, r_d and μ with I_D.

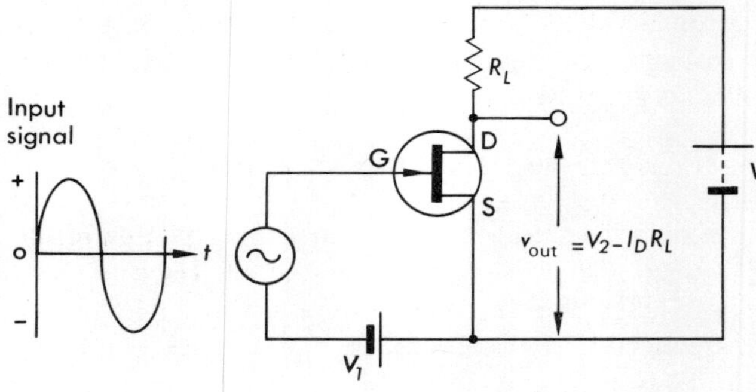

(a)

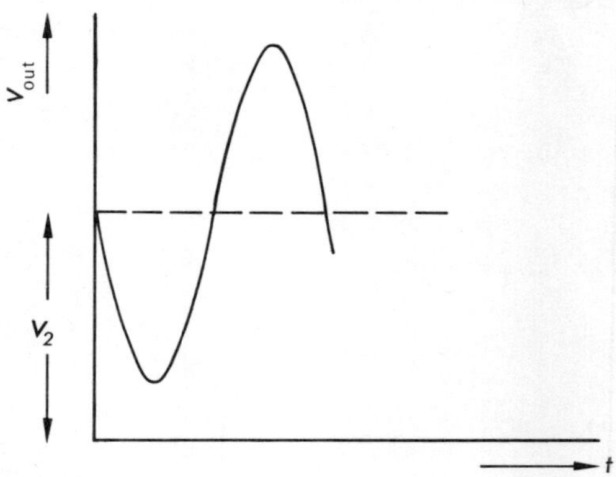

(b)

Fig. 4.9 Simple FET amplifier.
(a) Circuit.
(b) Output voltage waveform.

4.7 Amplification

Fig. 4.9 shows a simple FET amplifier. The d.c. voltage level V_1, applied between gate and source, reverse-biases the pn junction and establishes a depletion layer which extends into the n-type channel. At the same time, the d.c. voltage V_2 augments the depletion layer and also causes a steady drain current to flow.

When an alternating input voltage is applied between the gate and the source, it is superimposed on the existing d.c. voltage. The result is that the effective reverse-bias voltage across the pn junction causes a corresponding fluctuation in the depletion layer width and the output current I_D. This alternating output current produces an alternating output voltage across the load resistor R_L which is an amplified replica of the input signal.

It is worth while noting that whilst the current in a bipolar transistor has to cross two pn junctions, i.e. emitter–base and base–collector, the current in a FET passes along a p-type or n-type channel. Furthermore the current is made up almost entirely of majority carriers (electrons for n-type channel, holes for p-type channel). There will of course be some minority carriers due to thermal generation of electron-hole pairs, but the ratio of majority carriers to minority carriers is so large that the current may be considered to be carried by majority carriers only. This leads to the FET being referred to as a majority carrier device.

One of the major advantages that the FET has over the bipolar transistor is that it is inherently less noisy. This is primarily due to

(a) the source-to-drain current crosses no pn junctions,
(b) the FET is a unipolar device.

By comparison, in a bipolar transistor most of the unwanted electrical noise (spurious electrical impulses) is developed by the main current having to cross two pn junctions and by recombination between majority and minority carriers.

4.8 Equivalent circuit

Fig. 4.10 shows two very widely used equivalent circuits of the FET; Fig. 4.10(c) may be seen to be very similar to the h-parameter equivalent circuit of the bipolar transistor. This is not surprising since it is obtained in much the same way.

The equivalent circuit is built up as follows:

(a) Between the gate and the source of the actual physical device there exists a reverse biased pn junction, and the depletion capacitance associated with this junction is represented in the equivalent circuit by the capacitor C_{gs}. The actual magnitude of C_{gs} is a function of the area of the junction, consequently it varies with the geometry of the device and may be anything from 5 to 30 pF.

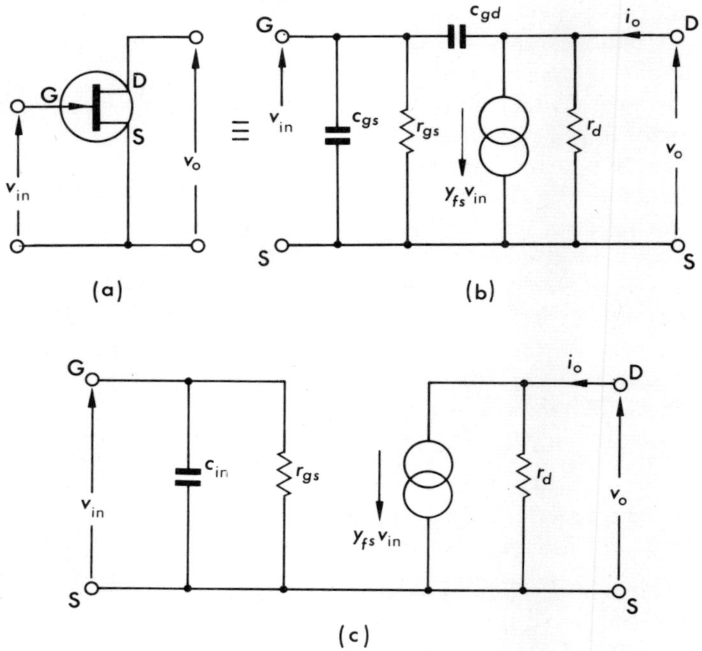

Fig. 4.10 Equivalent circuit for a junction FET.
(a) Device.
(b) Equivalent circuit.
(c) Simplified version of (b).

(b) Since the input signal is applied between the gate and source terminals, and since between these two terminals there exists a reverse biased pn junction, it follows that the input resistance of the device is very high. This input resistance is represented by the resistor r_{gs}, which may be typically 1000 MΩ.

Since the input circuit consists of a resistor in parallel with a capacitor, the actual input impedance is a function of frequency. Thus as the frequency of operation increases the input impedance decreases.

(c) Not only is the gate isolated from the source by a pn junction but it is also isolated from the drain. The depletion capacitance between gate and drain is represented by the capacitor C_{gd}. Fortunately the magnitude of this capacitance, which provides a capacitive path from output to input circuits, is small (due to the relatively large depletion layer), typically less then 5 pF.

(d) The current generator, connected between the source and the drain of the equivalent circuit represents the amplification of the device. The generator supplies a current to the output circuit which is controlled by the device's input voltage v_{in} and the magnitude of the supplied current is $y_{fs} v_{\text{in}}$, where y_{fs} is the forward transadmittance of the device.

(e) The resistor r_d represents the output or drain resistance which is high, typically between 100 kΩ to 1 MΩ.

In Fig. 4.10, the equivalent circuit (c) is a simplified version of circuit (b), which is obtained by referring C_{gd} to the input port of the circuit, i.e. C_{gd} appears between the input and output of the circuit and consequently it is modified by the Miller effect. In other words, C_{gd} is magnified and appears in the input circuit in parallel with C_{gs}, as a capacitor of value $(1-A)C_{gd}$, where A is the voltage gain of the stage (and for this connection will be negative). The total input capacitance C_{in} then appears to be $C_{gs}+(1+A)C_{gd}$. Usually manufacturers quote the input capacitance of a junction FET as C_{iss}. This represents the input capacitance with the output short circuited, in which case A would be zero, and for that condition,

$$C_{\text{in}} = C_{iss} = C_{gs}+C_{gd}.$$

4.9 Temperature Effects

Like all semiconductor devices, the junction FET suffers adversely when subjected to temperature variations. As the temperature increases the leakage current of the device increases, but more important than this are the considerable changes which occur in the d.c. parameters that lead to circuit design problems and to a limited working temperature range for the device. Temperature dependence of the FET parameters is attributed to different factors from those which govern the temperature dependence of bipolar transistors. For example, thermal runaway, a troublesome property of bipolar transistors, is not encountered with FETS.

4.9.1 *Leakage current*

The gate–channel junction, like any other reverse-biased semiconductor junction, will have a small leakage current. This gate leakage current has two components:

(a) gate-to-source leakage current,
(b) gate-to-drain leakage current.

Since the voltage between drain and gate is greater than that between source and gate, it follows that the gate-to-drain leakage current is the larger. However, although the total effect approximately doubles for every 8–10°C rise in temperature, the value of the leakage current at 25°C is only in the order of 1 nA. Consequently it is not too troublesome, except in certain bias circuits.

Some manufacturers specify the gate leakage current as I_{GDO}, I_{GSO}, and others as I_{GSS}. Fig. 4.11 indicates just how these measurements are made for an n-channel junction FET. In Fig. 4.11(a) it may be seen that I_{GDO} represents the gate leakage current which flows between drain and gate with the third terminal, the source, open. A small contribution to this current will still be made by the leakage current through the high resistance channel from the source end of the gate–channel junction.

By exchanging the source and drain connections in the above measurement of I_{GDO}, the leakage current due primarily to the reverse current between gate and source may be measured, i.e. I_{GSO}. Once again this leakage current will be supplemented by a component current which flows through the high channel resistance from the drain end of the gate–channel junction (Fig. 4.11b).

Finally, I_{GSS} represents the total gate leakage current which flows between gate and the source, with the drain shorted to the source. Note that this measurement includes the reverse current from the both ends of the channel, without either of them having to flow the length of the high resistance channel. In practice I_{GSS} is slightly less than the sum of I_{GDO} and I_{GSO} (Fig. 4.11c).

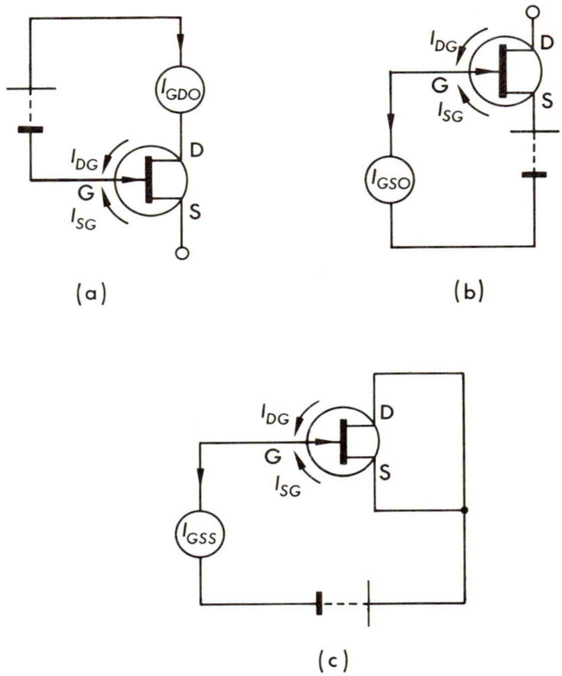

Fig. 4.11 Measurement of gate leakage current
(a) Measuring I_{DGO}.
(b) Measuring I_{SGO}.
(c) Measuring I_{GSS}.

4.9.2 *Variation of drain current with temperature*

There are two fundamental factors which control the variation of I_D with the temperature:

(a) Mobility of majority carriers in the channel.
(b) Variation of the gate-to-channel barrier.

The first factor predominates at higher values of I_D and actually causes I_D to decrease with temperature increase. This is, of course, opposite to the effect which an increase in temperature has on collector current in a bipolar transistor. The principal reason for this contrary behaviour is that the current in a bipolar transistor is carried by both minority and majority carriers (bicarrier or bipolar). The result being that as the temperature increases, the total current increases due to the exponential increase of minority carriers caused by electron–hole pair generation. On the other hand FETs are considered to be majority carrier devices. Although a minority carrier component is present it is considered negligible, so that the FET is a majority carrier or unipolar device. In consequence, as the temperature increases current flow through the device decreases.

This effect may be explained as follows: as the temperature of p- or n-type silicon *used in* FET *fabrication increases its resistivity increases,* even though there is an abundance of current carriers available within the material. This increase of resistivity with temperature is caused by the reduction of current carrier mobility, i.e. as the temperature increases the thermal energy of the lattice atoms increases, causing them to vibrate more vigorously about their mean positions. This makes collisions of current carriers with atoms more likely, which means that the carriers cannot move so freely through the material. Their mean free path is reduced which in turn decreases their mobility, resulting in a reduction of current through the material for a given applied voltage. An indication of the variation of the number of carriers available for conduction, the carrier's mobility, and the resistivity of the silicon used in FET manufacture, is given in Fig. 4.12.

It has been found that I_D decreases at a rate of 0·7%/°C rise in temperature at 25°C. This implies that temperature increase does not have the same disastrous effect on FETs as it does in bipolar transistors, i.e. thermal runaway is not encountered.

The second factor which controls the variation of I_D with temperature change is the variation of the gate-to-channel barrier or contact potential, i.e. as the temperature increases the width of the channel depletion layer decreases for a constant value of gate voltage. This permits an increase in channel current I_D. This is the same effect and of the same magnitude as the temperature variation of forward voltage in a bipolar transistor. It has been found that the resulting increase in I_D is equivalent to the gate–source reverse bias V_{GS} decreasing by approximately 2·2 mV/°C rise in temperature.

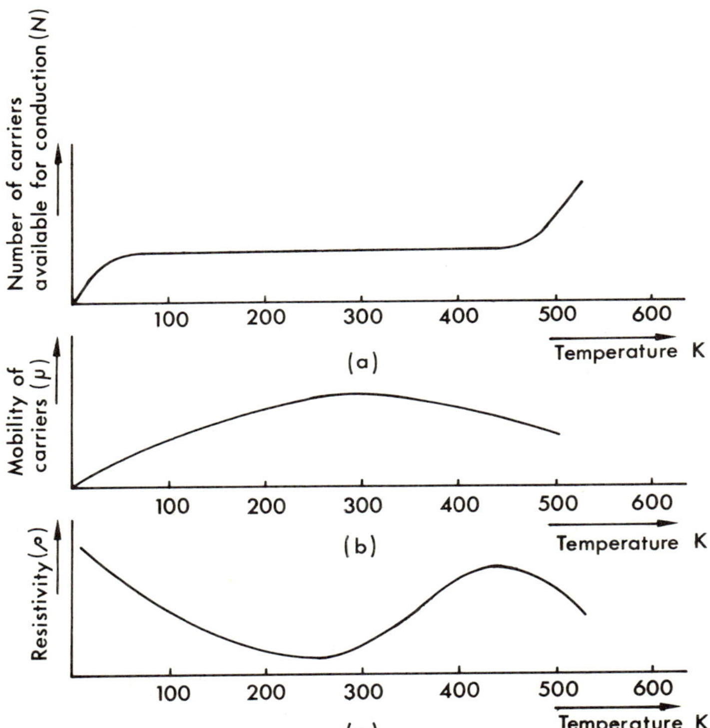

Fig. 4.12 Variation of N, μ and ρ with temperature for silicon.
(a) Variation in the number of current carriers available for conduction.
(b) Variation in the mobility of carriers.
(c) Variation in the resistivity of the silicon.

Note that since this increase in I_D with temperature is equivalent to the gate–source bias changing, it follows that if $g_m = \Delta I_D/\Delta V_{GS}$ and $\Delta V_{GS} = -2{\cdot}2$ mV/°C, the change in I_D due to channel barrier variation is a function of g_m.

Fig. 4.13 shows just how the above two effects influence the transfer characteristic of a junction FET. Note that I_D varies for every possible value of V_{GS} except that corresponding to point X.

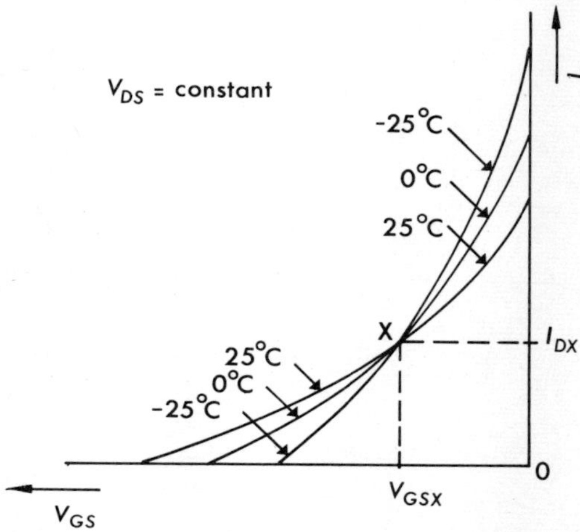

Fig. 4.13 The effect of temperature on the transfer characteristic

For drain currents above I_{DX}, I_D decreases with temperature increase and the material has a **positive temperature coefficient** of resistance. For values of I_D below I_{DX} the drain current increases with temperature increase, and thus the material has a **negative temperature coefficient** of resistance. At point X, however, I_D remains constant in spite of a considerable temperature variation: consequently point X is referred to as the **zero temperature coefficient point.**

Since I_D decreases at a rate given by 0·7 %/°C and since a change in gate voltage causes a change in I_D of $g_m \Delta V_{GS}$, the two effects are equal and opposite at point X, e.g.

$$0{\cdot}007\,|I_D| = 0{\cdot}0022\, g_m. \qquad [4.6]$$

Since from equation (4.1),

$$I_D = I_{DSS}\left[1-\frac{V_{GS}}{V_P}\right]^2$$

and from equation (4.2),

$$g_m = \frac{2I_{DSS}}{V_P}\left[1-\frac{V_{GS}}{V_P}\right],$$

on substitution in equation (4.6),

$$0{\cdot}007\, I_{DSS}\left[1-\frac{V_{GS}}{V_P}\right]^2 = 0{\cdot}0022\,.\,\frac{2I_{DSS}}{V_P}\left[1-\frac{V_{GS}}{V_P}\right].$$

This gives theoretical values of the gate-to-source bias voltage, V_{GSX}, for zero drift of I_D with temperature variation:

$$V_{GSX} = V_P+0{\cdot}64 \qquad \text{for n-channel FETS}$$

$$V_{GSX} = V_P-0{\cdot}64 \qquad \text{for p-channel FETS.}$$

4.10 Electrical Parameters and Maximum Ratings

To use the junction FET as a circuit component one must understand not only the theory of operation of the device but also the data which manufacturers supply with their devices. Unfortunately the junction FET, like all other semiconductor devices, has been given different labels for the same parameters and ratings. In the interest of standardization, the symbols utilized in Tables 4.1 and 4.2 are the Standard Letter Symbols for Semiconductor Devices recommended by the British Standards Institution.

The numerical values given to the few major parameters and and maximum ratings listed in Tables 4.1 and 4.2 may be considered to be typical for a general-purpose junction FET.

Note that in Table 4.1 symbols with triple subscripts are frequently employed. The first subscript indicates the object terminal, the second subscript indicates the common terminal, i.e. common source, common drain or common gate, and the third subscript gives the condition of the remaining terminal with respect to the common terminal. For example, I_{DSS} represents the saturated drain current of the common-source connected FET, whose gate is short circuited to the source. In a similar way, C_{iss} represents the input capacitance, that is the capacitance between input terminal and source with the third terminal (drain) short circuited to the source.

In Table 4.2, the maximum voltage ratings indicate the minimum terminal voltages that may be applied between any two terminals

TABLE 4.1 Major parameters of a junction FET

Parameter	Symbol	Min.	Typ.	Max.	Unit	Test condition
Maximum Source–drain *current*	I_{DSS}	10	15	30	mA	$V_{DS} = 15$ V, $V_{GS} = 0$
Pinch-off voltage	V	4	6	8	V	$V_{DS} = 15$ V, $I_D = 1$ nA
Forward transadmittance	y_{fs}	750	2500	5000	µS	$V_{DS} = 15$ V, $V_{GS} = 0$, $f = 1$ kHz
Input capacitance	C_{iss}		10	20	pF	$V_{DS} = 15$ V, $V_{GS} = 0$, $f = 1$ MHz
Transfer capacitance	C_{rss}		4	7	pF	$V_{DS} = 15$ V, $V_{GS} = 0$, $f = 1$ MHz
Leakage current	I_{GSS}		0·5	2	nA	$V_{GS} = 15$ V, $V_{DS} = 0$
Source–drain 'on' resistance	$r_{ds(on)}$		50	1000	Ω	$I_D = 0$, $V_{GS} = 0$, $f = 1$ kHz

which will lead to avalanche breakdown. There are two other widely-used symbols for expressing breakdown voltages, both of which use the previously mentioned triple subscript notation. For example, the maximum drain-to-gate voltage may frequently be seen expressed as BV_{DGO} or BRV_{DGS} and both are self-explanatory.

'On' resistance ($r_{ds(on)}$). This represents the resistance of the channel between the source and the drain for $V_{DS} < V_P$. This is an important parameter if the FET is being used for switching purposes since it affects the switching speed and the output level.

TABLE 4.2 Major maximum ratings of a junction FET

Maximum rating	Symbol	Value	Unit
Drain-to-source voltage	$V_{BR(DSO)}$	30	V
Drain-to-gate voltage	$V_{BR(DGO)}$	30	V
Gate-to-source voltage	$V_{BR(GSO)}$	30	V
Power dissipation	P_{tot}	300	mW
Junction temperature	T_j(max)	200	°C

It is measured by short circuiting the gate and source and by applying a small a.c. voltage between the source and the drain. The magnitude of the a.c. voltage should be small in order to ensure that the FET is not operating in the pinch-off condition. The circuit arrangement is illustrated in Fig. 4.14. The ratio of drain-to-source

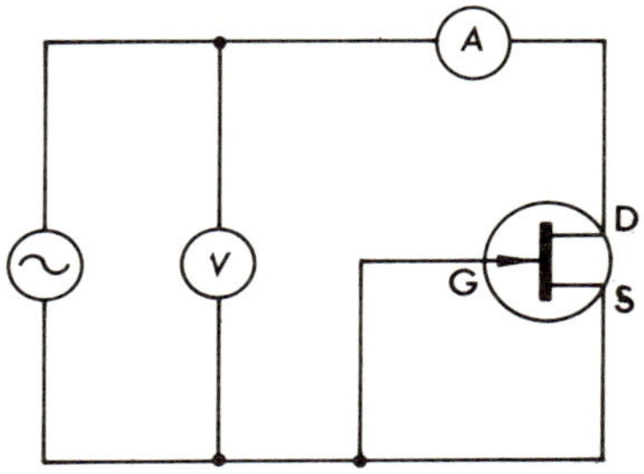

Fig. 4.14 Measurement of r_{ds} (on).

voltage v_{ds} to drain current i_d gives the 'on' resistance of the device, i.e.

$$r_{ds(on)} = v_{ds}/i_d$$

which is typically 50–1000 Ω.

Note: This is an a.c. measurement, with no d.c. voltages or currents involved.

Leakage Current (I_{GSS}). The magnitude of I_{GSS}, the leakage current of the reverse biased gate–source pn junction with the drain short circuited to the source, is an indication of the order of magnitude of the d.c. input resistance of the FET, i.e. the d.c. input resistance of a FET device is limited primarily by the gate leakage current. Because this leakage current is so temperature sensitive, e.g. at 25°C it may be, say, 2 nA while at 150°C it may be 5000 nA, it follows that the input resistance of the junction FET decreases as the temperature increases. In fact, the variation of I_{GSS} is such that it is usually specified at two different temperatures, e.g. 25°C and 100°C.

4.11 Junction FETs in Practice

The junction FET has three terminals (excluding the connection to the substrate) and consequently it may, like the bipolar transistor, be connected into a circuit in three possible ways. The connections are called **common-source** (CS), **common-gate** (CG) and **common-drain** (CD) (Fig. 4.15). The CG connection has a low input impedance; consequently it offers no real advantage over bipolar transistors, and as it is seldom used it will not be considered further.

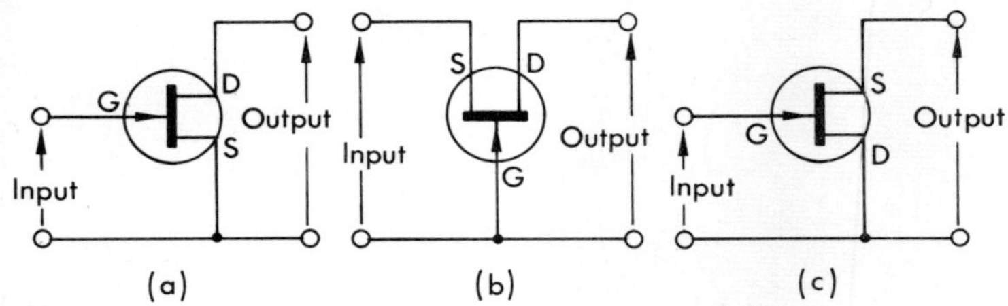

Fig. 4.15 The three junction FET configurations.
(a) Common-source.
(b) Common-gate.
(c) Common-drain.

4.11.1 *Common-source amplifier*

This connection, which is analogous to the common-emitter connection of a bipolar transistor, is very widely used. Its major advantages over the bipolar device are high input impedance and low noise. Since the amplification of a signal by a FET connected in the CS mode has already been considered, other practical aspects will be briefly examined in this section.

4.11.2 *Common-source biasing*

In order that any active device may operate effectively it is necessary to bias it correctly. Fig. 4.16 shows the simplest possible self-biasing circuit for a CS amplifier using an n-channel FET.

For correct operation of this n-channel FET it is necessary for:

(a) the drain to be positive with respect to the source,
(b) the gate to be negative with respect to the source.

This is readily achieved with a single supply voltage V_{DD} and a circuit connected as shown in Fig. 4.16. The drain current I_D flowing from drain-to-source through R_L and R_S ensures that the drain is positive with respect to the source. Thus the first biasing requirement is met. The second requirement is achieved as follows:

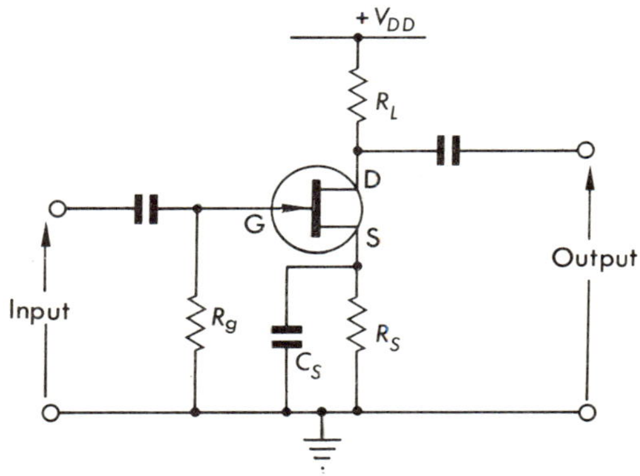

Fig. 4.16 Simple self-biased CS amplifier.

I_D flowing through R_S makes the source terminal of the FET positive with respect to earth. Now the gate current is negligible since only the leakage current flows through R_g and consequently the gate terminal is effectively at zero or earth potential. Thus if the source is positive with respect to earth and the gate is at earth potential, then the source must be positive with respect to the gate. In other words the gate is biased negatively with respect to the source.

As a numerical example concerning the biasing of the circuit shown in Fig. 4.16, let the FET used have a maximum drain current I_{DSS} of 2·0 mA and a pinch-off voltage V_D of 2·0 V. It is required to bias the circuit at a drain current I_D of say 1·0 mA. The correct value of gate bias V_{GS}, the self-bias resistor R_S and the operating value of g_m may be calculated as follows:

Substituting values in equation (4.1),

$$I_D = I_{DSS}\left[1-\frac{V_{GS}}{V_P}\right]^2,$$

$$1.10^{-3} = 2.10^{-3}(1-\tfrac{1}{2}V_{GS})^2,$$

$$\sqrt{0{\cdot}50} = 1-\tfrac{1}{2}V_{GS},$$

therefore $$V_{GS} = 0{\cdot}586\ V.$$

Since $$V_{GS} = I_D R_S,$$

$$R_S = V_{GS}/I_D$$

$$= 0{\cdot}586/1.10^{-3} = 586\ \Omega.$$

The operating value of g_m may be calculated from equation (4.2),

$$g_m = \frac{2.I_{DSS}}{V}\left[1 - \frac{V_{GS}}{V}\right]$$

$$= \frac{2.2.10^{-3}}{2}\left[1-\frac{0{\cdot}586}{2}\right]$$

$$= 2.10^{-3}.0{\cdot}707$$

$$= 1{\cdot}4\ \text{mA/V}.$$

The capacitor C_S which is connected across R_S has a reactance which may be neglected at all but the lowest signal frequencies.

Although the FET has a high input impedance Z_{FET} at low frequencies, when connected in an amplifier such as that shown in Fig. 4.16, the necessary inclusion of R_g gives an effective input impedance of Z_{FET} in parallel with R_g, the result of which will always be less than R_g. Since a major advantage of the FET over the bipolar transistor is its high input impedance, it is obvious that in order to maintain this advantage, R_g must be made as large as possible. Unfortunately however, due to the fact that the total gate leakage current must pass through R_g, too high a value of R_g will cause the gate bias voltage to vary too much with temperature. In practice then the value of R_g must be a compromise. (The actual function of R_g is to provide a path for the gate leakage current.)

As an example of how the gate bias voltage will vary with temperature, consider a temperature increase from 25°C to 50°C; let R_g be 10 MΩ and I_{DGO} be 5·0 nA at 25°C. Since the leakage current of a silicon device approximately doubles for every 8°C rise in temperature, the leakage current at 50°C will be approximately 40 nA, i.e.

$$\Delta I_{DGO} = (40-5)\times 10^{-9}\ \text{A} = 35\ \text{nA}.$$

The resulting change in gate potential ΔV_G will be

$$\begin{aligned}\Delta V_G &= \Delta I_{DGO} R_g \\ &= 35 \,.\, 10^{-9} \times 10 \,.\, 10^6\,\text{V} \\ &= 0{\cdot}35\ \text{V}.\end{aligned}$$

Since under normal operating conditions the drain current is dependent on the gate voltage, variation of gate bias results in a considerable change in drain current, i.e. the quiescent or no-signal operating point of the circuit is shifted. This may lead to distortion of the output voltage due to peak clipping, as shown in Fig. 4.17.

Variation of the gate leakage current with temperature is not the only consideration for correct biasing. Another factor is the spread

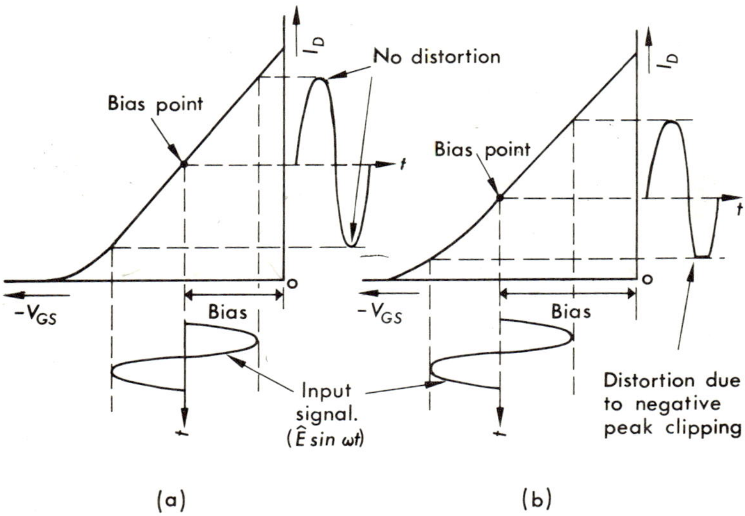

Fig. 4.17 Example of negative peak clipping due to increases in bias.
(a) No distortion.
(b) Distortion due to clipping.

of characteristics between devices of the same type number. Although bipolar transistors also suffer from spread in characteristics, the variation is not so great as with FETs. A typical example of how the transfer characteristic of a FET may vary due to extreme temperature and device variation is given in Fig. 4.18.

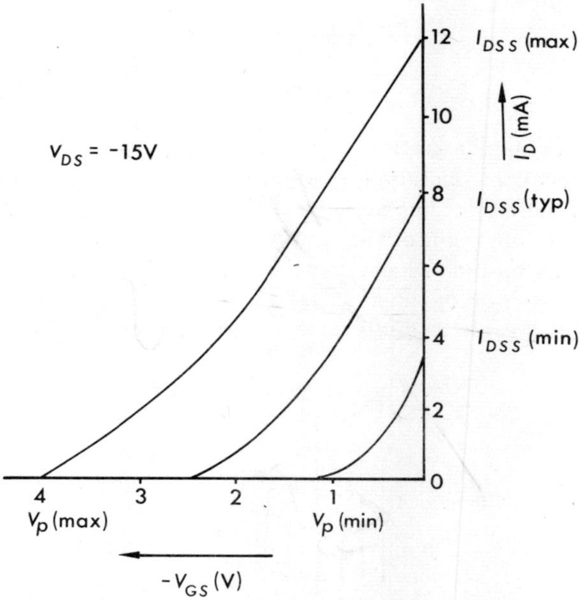

Fig. 4.18 Extreme variation of the transfer characteristic of a junction FET.

A circuit giving much better bias stabilization is shown in Fig. 4.19. The actual gate-to-source bias is provided by potential divider R_1, R_2 and the source resistance R_S. The gate potential V_G is defined by R_1 and R_2 and is given by

$$V_G = \frac{R_2}{R_1 + R_2} \cdot V_{DD}$$

whilst the source potential is defined by $I_D R_S$. Therefore the gate-to-source bias V_{GS} is given by

$$V_{GS} = V_G - I_D R_S .$$

ID = VG − VGS / RS

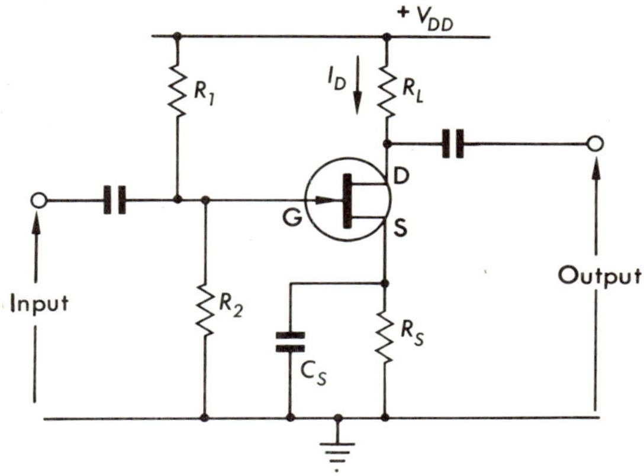

Fig. 4.19 Potential divider self-bias.

This straight line equation is useful in determining values of R_S and V_G for suitable d.c. biasing for a particular parameter spread. For example: it is required to bias a FET so that the maximum and minimum quiescent currents do not exceed I_D(max) and I_D(min) respectively. Points Q_1 and Q_2 are inserted on the maximum and minimum transfer curves as shown in Fig. 4.20. The bias line is then

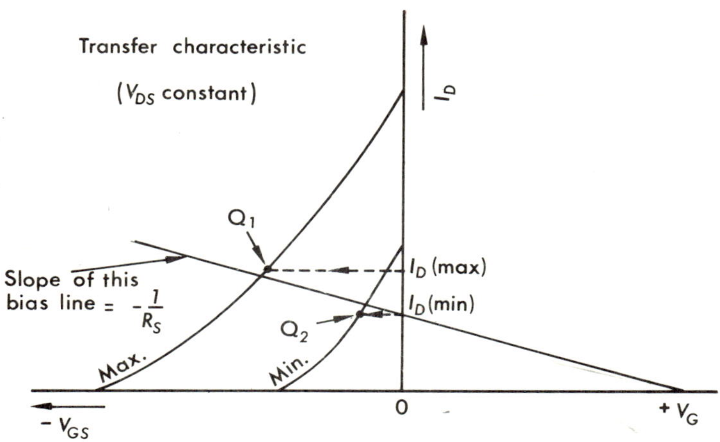

Fig. 4.20 Bias line for CS amplifier.

superimposed on the characteristic so that it passes approximately midway between points Q_1 and Q_2. The reciprocal of the resulting bias line gives the necessary value of R_S and the point of intersection of the bias line and the horizontal axis gives the corresponding value of the required gate voltage V_G. Since

$$V_G = \frac{R_2}{R_1 + R_2} \cdot V_{DD},$$

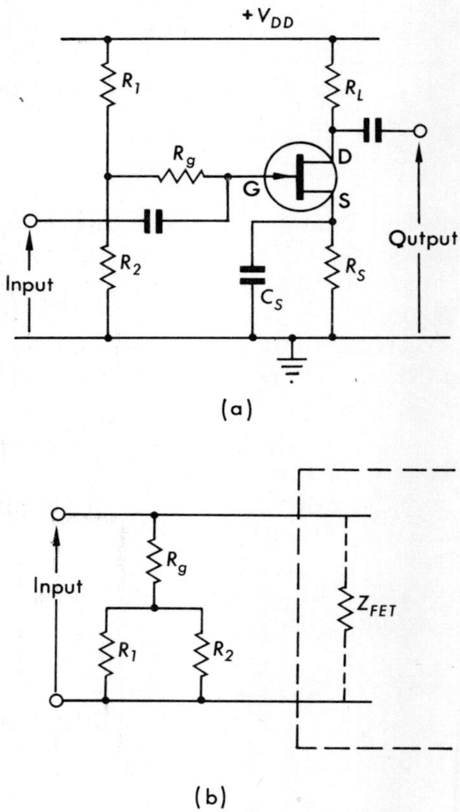

Fig. 4.21 CS amplifier with high input resistance and good stability.
(a) Circuit.
(b) Equivalent input circuit.

for a given supply voltage and a value of V_G specified by the bias line, suitable values of R_1 and R_2 may be determined. Note that since R_1 and R_2 effectively shunt the input of the amplifier, it is necessary for R_1 and R_2 to be as large as possible in order to maintain a high input impedance. Frequently in practice a fairly high resistor (1 – 5 MΩ) is connected between the junction of the potential divider and the gate, such as shown in Fig. 4.21(a). Note that the input signal is applied via the coupling capacitor directly to the gate terminal and not to the junction of the potential divider. As seen by the equivalent input circuit Fig. 4.21(b) the insertion of R_g effectively increases the input impedance of the amplifier from

$$\frac{R_1 R_2}{R_1 + R_2} \quad \text{in parallel with } Z_{\text{FET}}$$

to

$$R_g + \frac{R_1 R_2}{R_1 + R_2} \quad \text{in parallel with } Z_{\text{FET}}.$$

Providing R_g is not too large, the voltage drop across it is negligible and the bias line equation $V_{GS} = V_G - I_D R_S$ can still be used.

4.11.3 *Voltage gain of an audio-frequency (AF) CS amplifier*

Figure 4.22 shows an AF amplifier and its equivalent circuit. Note that the capacitors C_1, C_2, X_S, the resistor R_S and the supply voltage V_{DD} are not included in the equivalent circuit. This is because Fig. 4.22(b) is the equivalent circuit for alternating signals. In consequence it is assumed that the coupling capacitors C_1, C_2 and the bypass capacitor C_S have reactances which may be neglected at signal frequencies (R_s is short circuited by C_s) and the d.c. supply has negligible impedance. Furthermore, since it is an AF amplifier the parasitic capacitances C_{gd} and C_{gs} have very large reactances and are therefore assumed to have negligible shunting effect.

From the equivalent circuit, the output voltage v_{out} of the amplifier is given by

$$v_{\text{out}} = -y_{fs} v_{gs} \cdot \frac{R_L r_d}{R_L + r_d}.$$

The negative sign indicates 180° phase shift of the output voltage

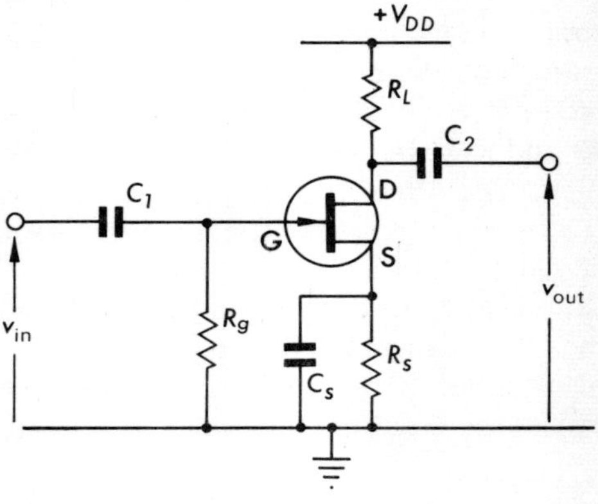

(a)

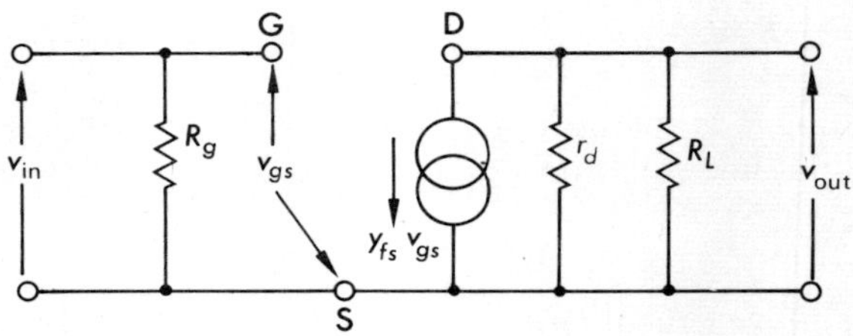

(b)

Fig. 4.22 AF amplifier and equivalent circuit.
(a) Circuit.
(b) Equivalent circuit.

waveform with respect to the input voltage. The voltage gain is

$$A_v = v_{\text{out}}/v_{in},$$

and, since $v_{in} = v_{gs}$, which is the signal voltage appearing between the gate and the source,

$$A_v = \frac{v_{\text{out}}}{v_{gs}}$$

$$= -y_{fs} \cdot \frac{R_L r_d}{R_L + r_d}.$$

The value of r_d for a junction FET is typically greater than 100 kΩ. If this is considerably greater than R_L, which is often the case, the voltage gain expression reduces to

$$A_v = -y_{fs} R_L .$$

Fig. 4.22(b) shows that at low frequencies the effective input impedance of the CS amplifier is R_g, ($r_{gs} \gg R_g$) and that the output impedance is simply the parallel combination of r_d and R_L. At very low signal frequencies, the increasing reactances of C_1, C_2 and C_s would cause the voltage gain to decrease from the above value of $y_{fs} R_L$.

4.11.4 *Common-drain amplifier or source-follower*

This connection is analogous to the common-collector or emitter-follower connection for the bipolar transistor and it has the same important properties:

(a) No phase reversal of the signal voltage.
(b) A voltage gain of less than unity.
(c) A high input impedance.
(d) A low output impedance.
(e) A high power gain.

Since one of the outstanding features of the FET is its inherently high input impedance, full advantage may be taken with this particular connection. In consequence, it is becoming very widely used as the first stage of an amplifier, where a very high input impedance is mandatory. It is also frequently employed for impedance transformation when FETs are used in the same circuits with bipolar transistors.

Fig. 4.23 shows a source-follower together with its a.c. equivalent circuits. The effective value of C_{in} of the simplified equivalent circuit is determined in exactly the same way as C_{in} was found for the CS connection, i.e. C_{gs} appears between the input and output circuits; it is therefore subjected to the Miller effect and its apparent value across the input terminals is $C_{gs}(1-A)$. Therefore the total input capacitance for the source-follower is given by

$$C_{\text{in}} = C_{gd} + C_{gs}(1-A)$$

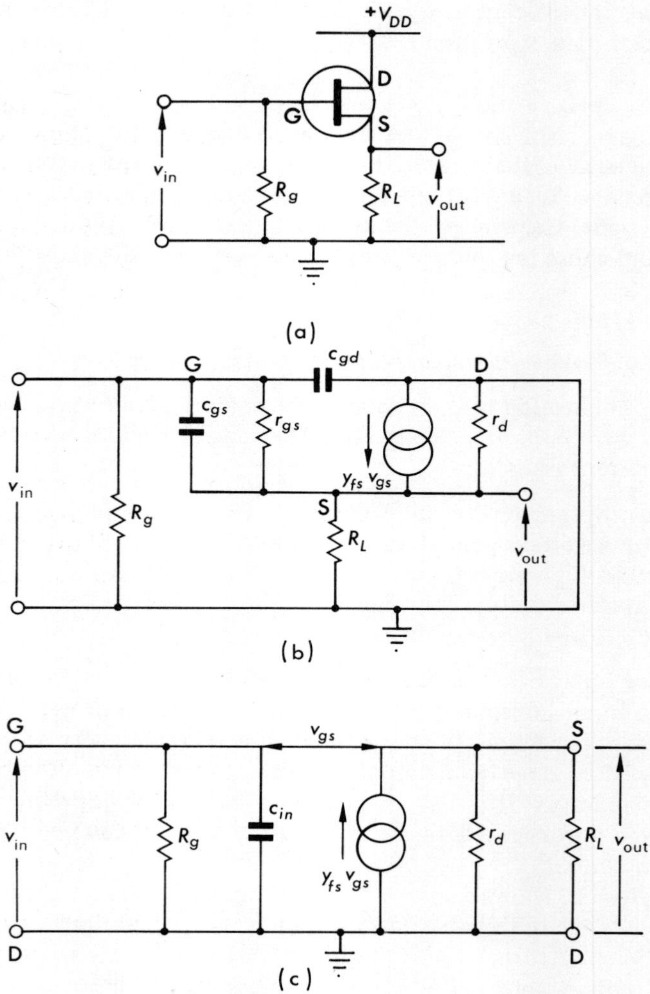

Fig. 4.23 Simple source-follower and equivalent circuits.
(a) Circuit.
(b) Equivalent circuit.
(c) Simplified version of (b).

Note that r_{gs} has been omitted from the simplified equivalent circuit; this is because the input impedance is effectively determined by C_{in}. In other words r_{gs}, which is typically 1000 MΩ, is so large that its shunting effect on C_{in} is negligible.

It is worth comparing the respective values of C_{in} for CS and CD connections. With the CS connection, the output voltage was not only considerably greater than the input voltage but it was phase inverted with respect to it. Thus the voltage gain (A_v) was greater then unity and negative. This resulted in a fairly high input capacitance even though the parasitic capacitances C_{gs} and C_{gd} were small, e.g. if $C_{gs}=20$ pF, $C_{gd}=5$ pF and $Av = 19$, the total input capacitance is given by:

$$\begin{aligned} C_{in} &= C_{gs}+C_{gd}(1-Av) \\ &= 20+5(1+19) \\ &= 120 \text{ pF.} \end{aligned}$$

Now with the CD connection, the output voltage is not only in phase with the input voltage but less than it, giving a voltage gain which is *positive* and less than unity. By using the above values for C_{gs} and C_{gd} and by letting $Av = 0{\cdot}95$ the total input capacitance is given by

$$\begin{aligned} C_{in} &= C_{gd}+C_{gs}(1-Av) \\ &= 5+20(1-0{\cdot}95) \\ &= 6 \text{ pF.} \end{aligned}$$

Hence the CD connection has a much lower input capacitance than the CS connection. Since the fall-off of input impedance with increase in frequency is primarily a function of C_{in}, it is obvious that the input impedance of the CD connection does not decrease so drastically as that of the CS connection. Furthermore, because the high-frequency response of a FET amplifier is limited by its decreasing input impedance, it follows that the frequency response of a source-follower is superior to that of a CS connection.

4.11.5 *Source-follower bootstrap circuit*

The necessary d.c. self-bias for a source-follower may be obtained very simply by a circuit such as that shown in Fig. 4.23(a) or, alternatively, in a similar way to that depicted for the CS amplifier in Figs. 4.19 and 4.21. Unfortunately, however, the overall input impedance of these circuits is limited by the presence of the gate-to-source d.c. path necessary for the FET's leakage current. Thus although the source-follower may have a higher input impedance than the CS connection, this valuable property will be somewhat nullified by the shunting effect of the d.c. bias network of the gate.

One method which overcomes this and is frequently employed for obtaining a high input impedance is a technique known as *bootstrapping*. The principle of this technique is to make the potentials at the opposite ends of a resistor rise and fall together, so that the actual potential difference between the two ends is small.

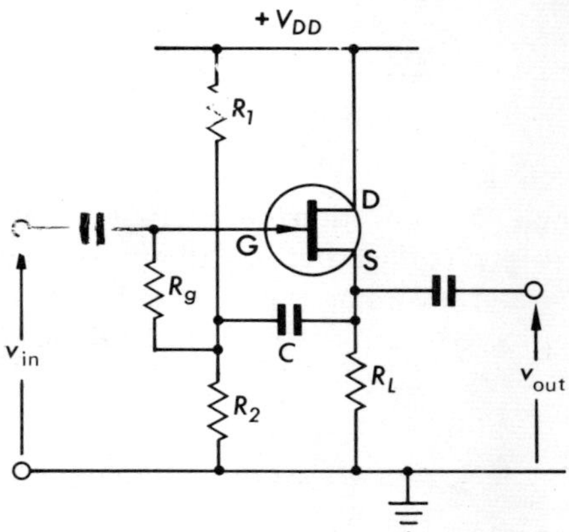

Fig. 4.24 Source-follower bootstrap circuit.

Fig. 4.24 shows a source-follower bootstrap circuit. The d.c. self-bias is provided by the source resistor R_L and the gate network consisting of R_1, R_2 and R_g in exactly the same way as that discussed for the CS amplifier. However, due to the bootstrapping, the effective input impedance 'seen' by the signal is considerably different from $R_g+R_1 R_2/(R_1+R_2)$.

Notice that the signal input voltage is applied via the coupling capacitor to the gate and the top end of R_g, whilst the bottom end of R_g is connected via a capacitor C to the source. Since the source voltage follows the gate voltage and is very nearly equal to it

(A_v approaches unity), then obviously as the voltage at the top end of R_g changes with the input voltage, the bottom end does so as well. Thus the a.c. potential across R_g is small, whilst the d.c. potential of the gate remains at a level determined by R_1 and R_2.

If the voltage gain of the source-follower is A_v, then since the whole of the output voltage is fed back via the capacitor to the bottom of R_g, the actual a.c. voltage v_1 across R_g must be given by

$$\begin{aligned} V_1 &= V_{in} - V_{out} \\ &= V_{in} - A_v V_{in} \\ &= V_{in}(1 - A_v) \end{aligned}$$

The a.c. current i_1 through R_g is given by

$$i_1 = \frac{V_1}{R_g} = \frac{V_{in}(1 - A_v)}{R_g}$$

Thus, the effective shunt impedance Z_1 of the gate bias network 'seen' by the signal must be

$$\begin{aligned} Z_1 &= \frac{v_{in}}{i_1} \\ &= \frac{v_{in}}{v_{in}(1 - A_v)/R_g} \\ &= \frac{R_g}{1 - A_v} \end{aligned}$$

Hence if $R_g = 2{\cdot}0$ MΩ and $A_v = 0{\cdot}95$, the apparent shunt resistance will be

$$\frac{2\,.\,10^6}{1 - 0{\cdot}95} = 40 \text{ M}\Omega$$

An example of just how effective bootstrapping can be in providing a high input impedance can be deduced from Fig. 4.25 which depicts a practical circuit with component values included. Without the 25 μF feedback capacitor, the approximate input impedance of this circuit is 2·2 MΩ shunted by 10 pF. However, with the capacitor connected, i.e. with the bootstrap circuit, the input impedance is raised to approximately 44 MΩ shunted by 10 pF—a substantial improvement for the addition of only one extra circuit component.

4.12 Applications of the Junction FET

Obviously, from what has already been said in this chapter, the junction FET is similar in some respects to a thermionic pentode valve and in other respects to a bipolar transistor. In fact it has been frequently said that a FET has the virtues of both without the vices of either. Unfortunately this is only partially true. Perhaps a better way of looking at the FET is to simply consider it as another semiconductor device that has some important attributes which make it particularly suitable for use in specific applications.

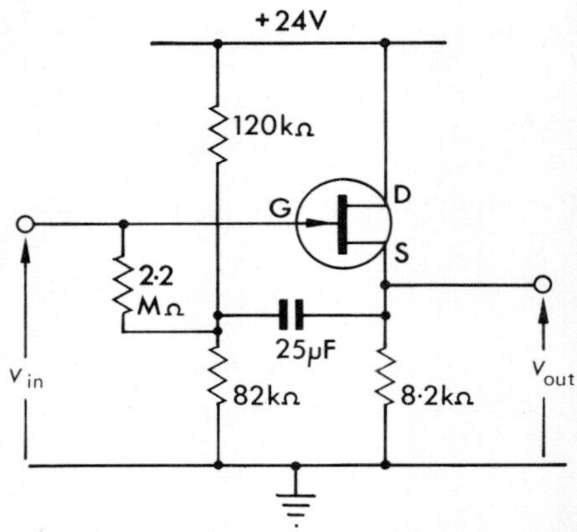

Fig. 4.25 A practical bootstrapped source-follower.

Most of the advantages and disadvantages of the FET have been already mentioned in this chapter and will not be considered further. It is, however, useful to list some of its properties which set it apart from other active semiconductor devices. They are:

(a) High input impedance.
(b) Voltage operated.
(c) Low noise.
(d) Controllable temperature coefficient.
(e) Voltage controlled resistance when $V_{DS} < V_P$.
(f) Very high 'on' to 'off' impedance ratio (switching applications).

Since the junction FET is a relatively new device, its development is not yet complete and in consequence the device suffers from a number of limitations such as

(a) Maximum frequency of operation is approximately 500 MHz.
(b) Total device dissipation relative to bipolar transistors is low.
(c) A limitation on their use is that currently they are more expensive than similarly rated bipolar transistors.

There is nothing in the physics of the junction FET to suggest that the above limitations are but temporary. With regard to which device to use for a particular application, the best criterion to adopt must be to employ the device which will give the desired results most cheaply and reliably.

4.12.1 *Compound or hybrid source-follower*

The two outst nding properties of the junction FET are its high impedance and low noise. These important attributes may be used to full advantage by making the junction FET the first stage of an amplifier. Now due to the contrasting properties of FETs and bipolar transistors, it is possible to obtain a higher overall power gain for an amplifier by using both FETs and bipolar transistors than it is by using either alone, e.g. for a given overall gain from a multistage amplifier, the FET operates as the first stage, but once the high impedance level has been reduced it is more economical to use bipolar transistors.

Fig. 4.26 shows one way in which the performance of a circuit may be improved by using both unipolar and bipolar transistors together. The two circuits are similar except that (b) utilizes a bipolar transistor instead of a resistor as the load of the FET. The effect of this loading is to present a source resistance which is equal to the large dynamic resistance seen looking into the collector of the transistor. The result of this very high dynamic load is a voltage gain of 0·99 compared to 0·95 for the resistor load source-follower of circuit (a). This considerable increase in voltage gain also results in a much higher input impedance, e.g. the effective value of the 2·2 MΩ

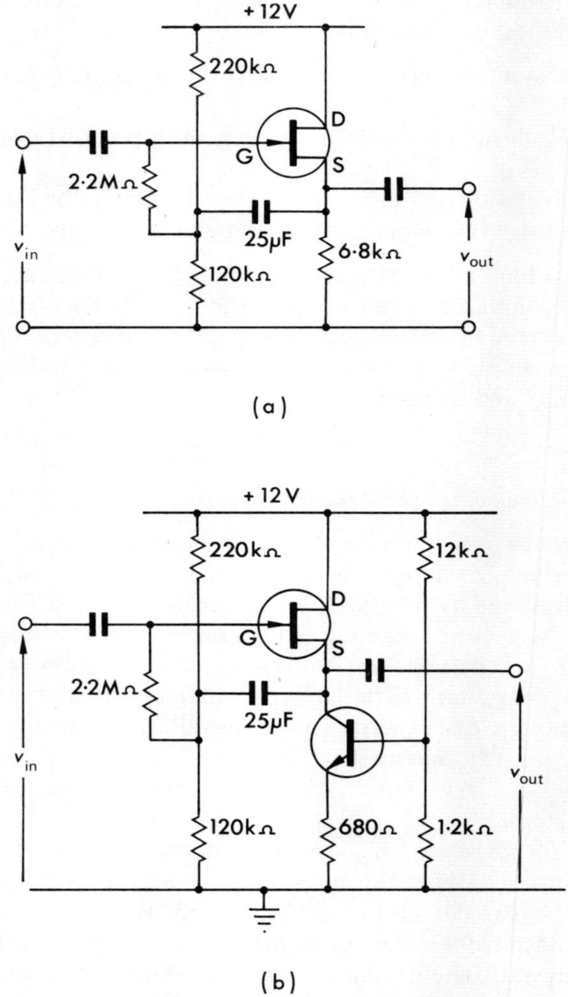

Fig. 4.26 Compound or hybrid source-follower.
(a) Source-follower, resistor loaded.
(b) Source-follower, transistor loaded.

gate resistor, say R_g', becomes

$$R_g' = \frac{R_g}{1 - A_v}$$

$$= \frac{2{\cdot}2\ \text{M}}{1-0{\cdot}99}$$

$$= 220\ \text{M}\Omega$$

So that if r_{gs} of the FET is 1000 MΩ, the approximate input resistance becomes

$$\frac{Rg'\,r_{gs}}{R_g'+r_{gs}} = \frac{220\,.\,10^6\,.\,1000\,.\,10^6}{1220\,.\,10^6}\ \text{M}\Omega$$

$$= 180\ \text{M}\Omega$$

resulting in an input impedance of 180 MΩ shunted by 10 pF, compared with an input impedance of circuit (a) of 44 MΩ shunted by 10 pF. It is not practical to replace the bipolar transistor by a resistor whose resistance is equal to the dynamic collector resistance of the transistor, owing to the excessive voltage drop and power dissipation which would result.

4.12.2 *A junction FET as voltage variable resistor*

For most linear applications the junction FET is operated in the constant current portion of its characteristic, i.e. for $V_{DS} > V_P$. It is, however, possible to utilize the pre-pinch-off region of the characteristic. It may be seen from the output characteristic depicted in Fig. 4.27 that for V_{DS} less than V_P the slope of the characteristic is initially linear and a function of V_{GS}. In other words, the resistance of the channel between source and drain, r_{DS}, is constant providing the gate voltage is maintained constant, and the drain voltage is less than and not too near the pinch-off value. Further, any increase in the reverse bias between gate and source results in a corresponding increase in the channel resistance. Hence by manufacturing junction FETs which have a high pinch-off voltage (V_P = 10–20 V) and by operating such a device in the pre-pinch-off region of its characteristic a **voltage variable resistor** (VVR) is obtained. An indication of how channel resistance r_{DS} varies with reverse gate bias V_{GS} is given in Fig. 4.28.

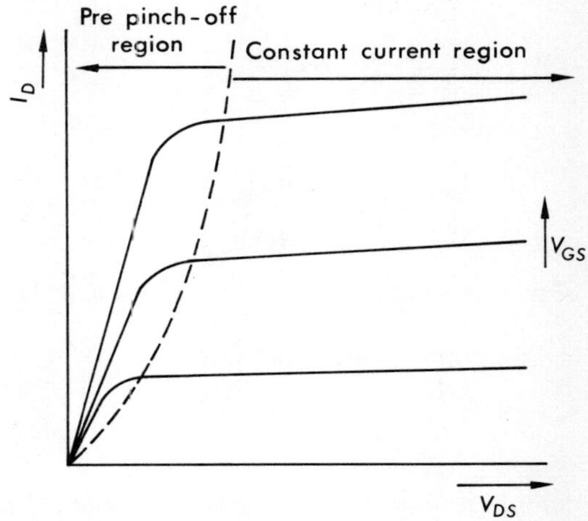

Fig. 4.27 Output characteristic of a junction FET.

The applications of junction FETs as VVRs are quite numerous—a typical example is shown in Fig. 4.29. Here, the VVR characteristic of a junction FET is used to automatically control the gain of an amplifier. Basically this control is obtained by using the FET as a

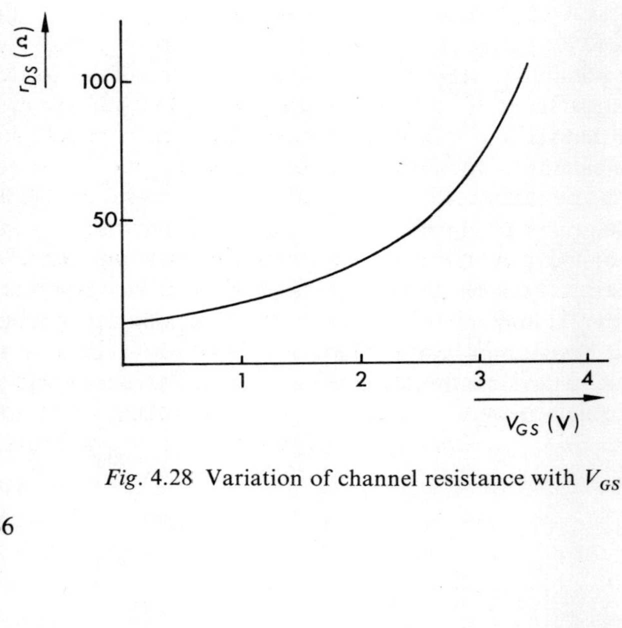

Fig. 4.28 Variation of channel resistance with V_{GS}.

variable emitter resistance for the first stage of the amplifier, i.e. variation of the FET's channel resistance varies the magnitude of the degeneration or negative feedback applied to the amplifier.

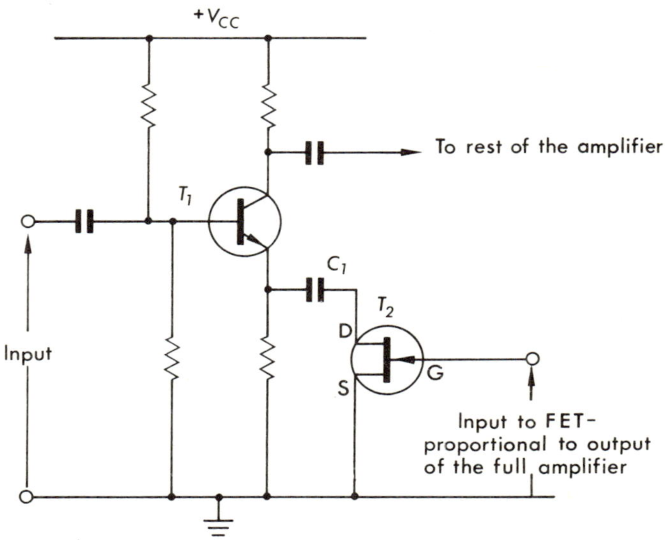

Fig. 4.29 Automatic gain control of an amplifier.

The amplified signal near the output of the full amplifier is first rectified and filtered so as to produce a d.c. voltage whose amplitude is proportional to the signal level. This voltage is applied to the gate of the FET which is connected in the emitter circuit of the first stage of the amplifier.

If the signal level causes the reverse bias of the FET to increase, the resistance source and drain increases causing the emitter impedance of T_1 to increase. This in turn produces a larger degenerative voltage at the emitter which decreases the gain of T_1 and therefore the rest of the amplifier, i.e. the original increase in output level is automatically offset.

Note: In order to prevent the automatic gain control signal from affecting the d.c. bias of T_1, the FET is d.c. isolated by C_1.

REFERENCES

EVANS, A. D. *Characteristics of Unipolar Field Transistors,* Electronic Industries, **22**, March 1963.

HOLCOMB, S. W. and L. J. SEVIN. *Field Effect Transistors for Low-Level Circuits*, Texas Instruments Ltd., 1968.

NOLL, E. M. *Field Effect Transistors; Principles, Experiments and Projects,* W. Foulsham & Co. Ltd., 1969.

SEVIN, L. J. *Field Effect Transistors*, McGraw-Hill Book Company, 1965.

SHOCKLEY, W. 'A Unipolar Field Effect Transistor', *Proc. Inst. Radio Engrs.,* **40**, 1365, 1952.

Chapter 5

Insulated-gate Field-effect Transistors

5.1 Introduction

Although the insulated-gate field-effect transistor, or IGFET, has only become commercially important since 1965, the principles of its operation had been thought of and patented prior to 1940, i.e. eight years before that of the bipolar transistor. Unfortunately, owing to fabrication difficulties, its development was retarded. Now however, semiconductor technology has advanced so rapidly that fabrication is no longer a problem and consequently IGFETs are extensively used in a wide variety of applications.

There are two basic forms of the IGFET, they are:

(a) **The Enhancement IGFET.**
(b) **The Depletion IGFET.**

These two forms may be subdivided, like the junction FET, into p- or n-type channel versions.

Although the principle of operation of both types of IGFET is similar, the principle does differ in a fundamental manner from that of the junction FET. The result is that both families of FETs have distinct advantages over each other and consequently they find use in differing applications.

Briefly, the essential difference is that with the junction FET the conductivity of the channel between source and drain is controlled by the transverse electric field created by the reverse-biased pn junction, whereas with both types of IGFET the conductivity of the channel is controlled by a transverse electric field which is induced capacitively across an insulator or dielectric.

5.2 Enhancement IGFET

Fig. 5.1 shows the essential features of an enhancement-type IGFET.

It consists of two relatively heavily doped p-type regions known as the **source** and the **drain** which are diffused into a n-type wafer of silicon, known as the **substrate**. Across the top of the device, excluding regions for connection to the source and drain, exists a film of silicon oxide. This silicon oxide film acts as an extremely good insulator between the surface of the substrate and the relatively

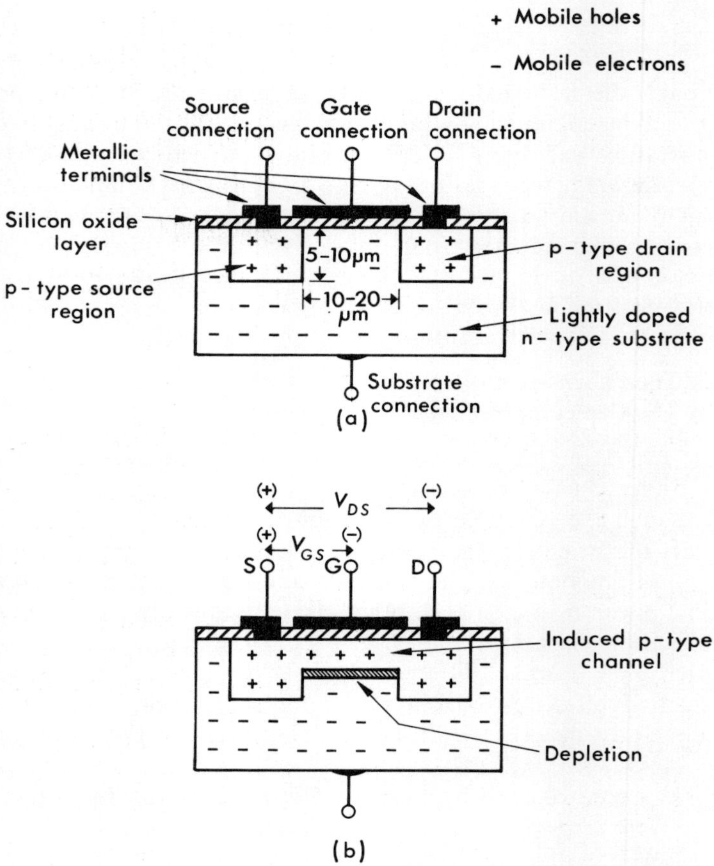

Fig. 5.1 Structure of an enhancement-type IGFET.
(a) No terminal voltages.
(b) Negative voltage applied to the gate.

large metallic layer to which the gate terminal is made. The insulating layer is the reason why the device is called the insulated-gate field-effect transistor. Note that although there is a source and a drain region, there is no gate region but merely a metallic layer. This metallic layer together with the insulating silicon oxide layer and the silicon substrate form a *capacitor*, i.e. the metallic layer and silicon substrate form the two 'conducting plates', with the silicon oxide acting as the dielectric.

When a negative potential is applied to the gate electrode (Fig. 5.1b) the metallic layer becomes negatively charged and the second plate of the capacitor, the surface of the substrate, becomes positively charged. It is this induction of charge in the material directly beneath the gate metallic layer and between the source and drain regions which forms the basis of the operation of both types of IGFET.

Frequently IGFETS are referred to as Metal Oxide Semiconductor Transistors or MOSTS. This name is derived from the structure of the device. From Fig. 5.1 it may be seen that the gate electrode is a metallic layer formed on silicon oxide, i.e. **M**etal on **O**xide of **S**ilicon.

5.2.1 *Operation of the enhancement* IGFET

Consider the enhancement IGFET shown in Fig. 5.1(a). Between source and drain there exist two pn junctions back to back, i.e. from source to drain the regions are doped respectively p–n–n–p. If a negative potential is applied to the drain with respect to the source, then only a very small current will flow. This is because the drain–channel pn junction will be reverse biased and consequently the only current which can cross this junction is its very small leakage current. In order to keep this leakage current small, the substrate is lightly doped compared with the relatively heavily doped source and drain regions.

If the substrate is earthed and a negative voltage applied to the gate, the resulting electric field penetrates the substrate region between the source and the drain and produces a two-fold effect:

(a) Electrons are repelled from the substrate region immediately below the gate terminal.
(b) Holes, which are the minority carriers within the substrate are attracted to the region vacated by the electrons.

The result is that the channel region directly beneath the gate electrode becomes rich in holes and so effectively changes from n- to p-type material (see Fig. 5.1b).

If the drain is made negative with respect to the source a current consisting of holes will flow through the p-type material. The magnitude of this current will be a function of V_{DS} and the gate bias voltage V_{GS}. (Frequently, since the substrate is only lightly doped, it is common practice to connect the substrate to the source, so that the gate attracts additional carriers from the source into the channel).

The required voltage necessary to change the conductivity of the channel region so that a drain current may flow is called the **threshold voltage** (V_T), and is typically 2–5 V, depending upon the doping of the substrate and thickness of the oxide.

This particular type of device is called a p-channel enhancement IGFET. This is because the conductivity between source and drain is improved or enhanced by the application of V_{GS} and the conductivity of the channel so formed is p-type.

With the junction FET the input resistance was high on account of the reverse biased pn junction between gate and channel. With the IGFET the input resistance is even higher, since the gate electrode is physically insulated by silicon oxide from the channel. It may be as high as 10^{15} ohms.

5.3 Depletion IGFET

Fig. 5.2(a) shows that basically the structure of the depletion-type IGFET is very similar to the enhancement type. The essential difference is the deliberate introduction during manufacture of impurities of the correct type so as to provide a ready-made lightly-doped channel, i.e. a low resistance channel exists between source and drain in the absence of terminal voltages. Note that all three regions—source, channel and drain—are of the same type of material.

5.3.1 *Operation of the depletion* IGFET

Consider the depletion-type IGFET shown in Fig. 5.2(b). Between the n-type source and drain regions is a lightly-doped n-type channel so that if the drain is made positive with respect to the source a current will flow. This occurs because, since there are no reverse-biased pn junctions, the current path is entirely through n-type material. Notice that this is opposite to the enhancement-type FET, and because of this the depletion-type IGFET is referred to as a **normally-on** device and the enhancement-type as a **normally-off** device.

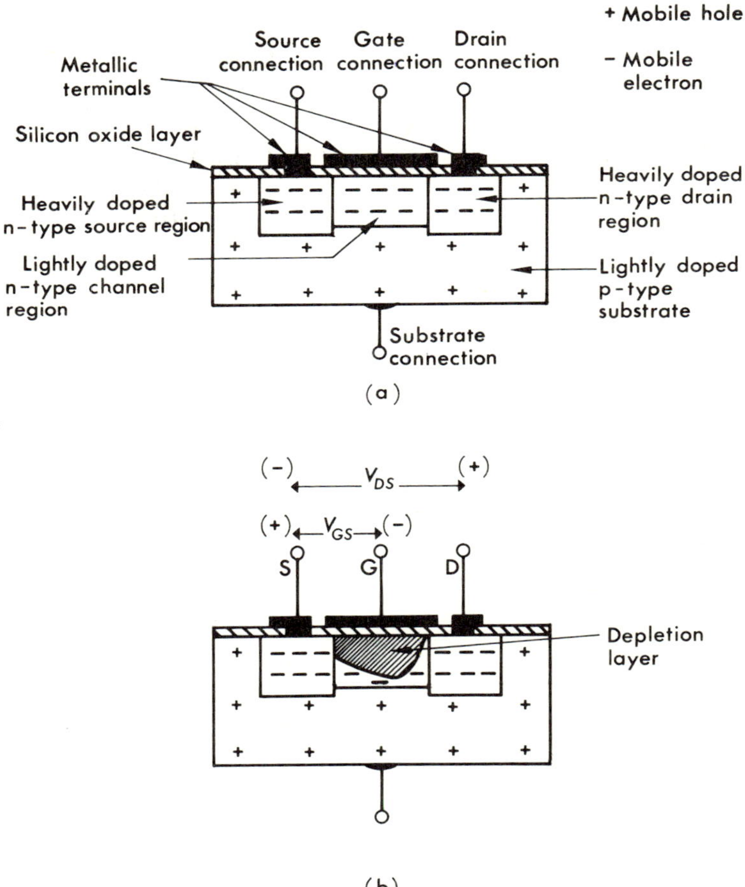

Fig. 5.2 Structure of a depletion-type IGFET.
(a) No terminal voltages.
(b) Negative voltage applied to the gate.

If the substrate is short circuited to the source and a voltage is applied to the gate so as to make it negative with respect to the source, the resulting electric field penetrates into the channel, causing electrons to be repelled from it and so making the channel less conductive. Thus since the resistivity of the channel region may be controlled by the magnitude of the gate bias voltage, the current flowing through the device is a function of both V_{GS} and V_{DS}, e.g. for a fixed value of V_{DS} the drain current may be reduced by increasing the negative gate bias voltage.

Although the enhancement-type IGFET can *only* be operated by making the gate potential either positive **or** negative, **depending on the type of doping of the channel.** The depletion-type IGFET may be operated by both positive and negative gate potentials **irrespective of channel doping.**

5.4 Effect of V_{DS} on the Channel Conductivity of an IGFET

Having briefly considered the structure and operation of both types of IGFET, it is necessary, in order to understand more fully their electrical characteristics, to examine the shape and conductivity of the channel region under the influence of a variable V_{DS}, whilst maintaining a constant gate bias V_{GS} (see Fig. 5.3).

Figure 5.3(a) depicts the shape of the channel region in the absence of V_{DS}, so that with the substrate short circuited to the source, the active channel region is rectangular. Now just as with the junction FET the shape of the depletion layer was a function of both V_{GS} and V_{DS}, so with the IGFET the active channel region is not only influenced by the transverse electric field produced by V_{GS}, but also by the potential gradient between the source and the drain. The result of this is that the active region of the channel becomes non-uniform, i.e. wedge shaped. For low values of V_{DS} the lateral effect is small and this is depicted by Fig. 5.3(b). If, however, V_{DS} is increased, the drain current increases, so causing the voltage drop along the channel length to increase, i.e. the potential gradient increases which causes the lateral effect on the active region to increase. The result is that the conductivity of the channel decreases as the drain end is approached. In other words the channel region is becoming **pinched off**. Figure 5.3(c) illustrates the condition when $V_{GS} = V_{DS}$. Note

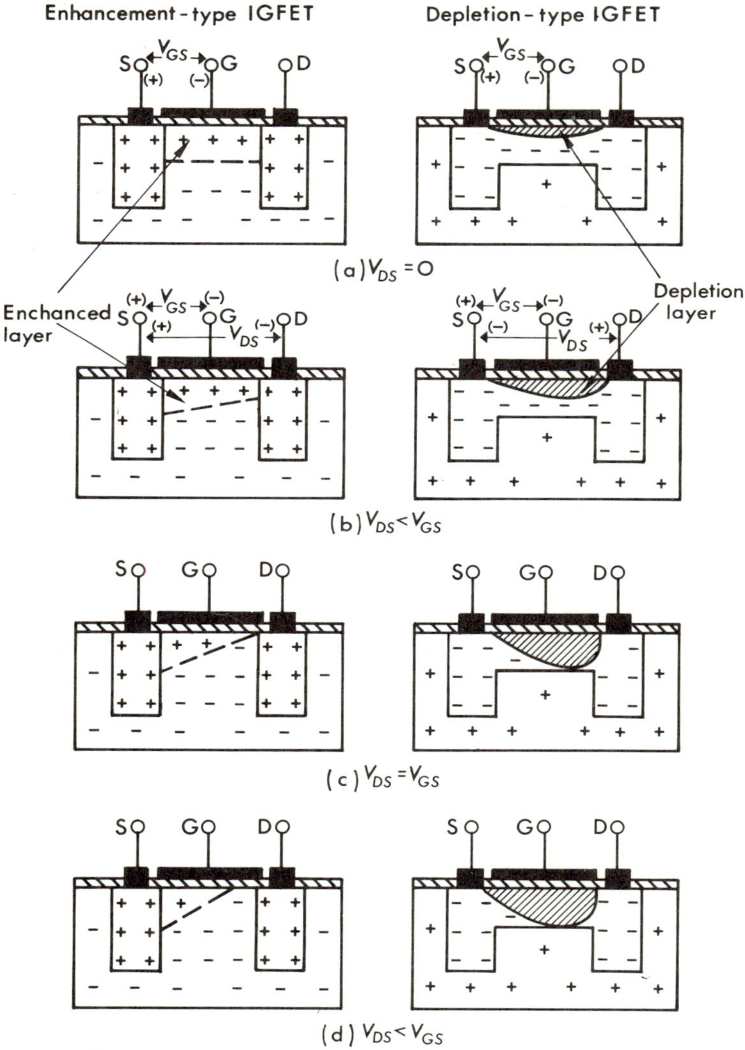

Fig. 5.3 Variation of channel shape with variation of V_{DS} (V_{GS} constant).
(a) $V_{DS} = 0$ (c) $V_{DS} = V_{GS}$
(b) $V_{DS} > V_{GS}$ (d) $V_{DS} > V_{GS}$

that if $V_{GS} = V_{DS}$ there can be no resulting electric field at the channel junction with the drain region.

If V_{DS} continues to increase ($V_{DS} > V_{GS}$) then the enhanced or depleted regions will cease to exist at some point before the channel–drain junction is reached. This condition is shown in Fig. 5.3(d). The actual point will be somewhere between the source and the drain where the channel potential (governed by V_{DS}) is equal to the gate potential V_{GS}.

The channel does not become completely pinched off under the condition that $V_{DS} > V_{GS}$. If it did, the drain current would cease, so causing the lateral effect to collapse which would result in the enhanced or depletion zones being re-established. In practice an equilibrium condition is maintained under which the drain current remains practically constant in spite of considerable variation of V_{DS}, i.e. the drain current reaches saturation and remains fairly constant until the magnitude of V_{DS} is such that avalanche breakdown occurs.

5.5 Symbols and Bias Polarities

Fig. 5.4 illustrates the symbols used for the four different types of IGFET. As would be expected the symbols used are very similar to those used for the junction FET. There must, however, be distinctions in order to represent the different devices unambiguously. The IGFET is distinguished from the junction FET by indicating the insulated nature of the gate, i.e. the gate electrode is shown remote from the other electrodes. A distinction between the depletion-type and the enhancement-type device is obtained by representing the channel of the former with a full line and the channel of the latter with a broken line. The possible reasoning behind this representation is the initial state of the channel of the two devices. The depletion-type IGFET is a 'normally-on' device, i.e. a channel exists between source and drain in the absence of a gate potential. The full line in the symbol may be considered as representing this existing channel between source and drain electrodes. On the other hand, the enhancement IGFET is a 'normally-off' device, i.e. the channel has to be formed by a gate potential before current can flow from source to drain. Thus the channel is incomplete and so is only represented by a broken line.

The p- and n-type channel devices are distinguished from each other by the direction of the arrow head—as was the case for junction FETS.

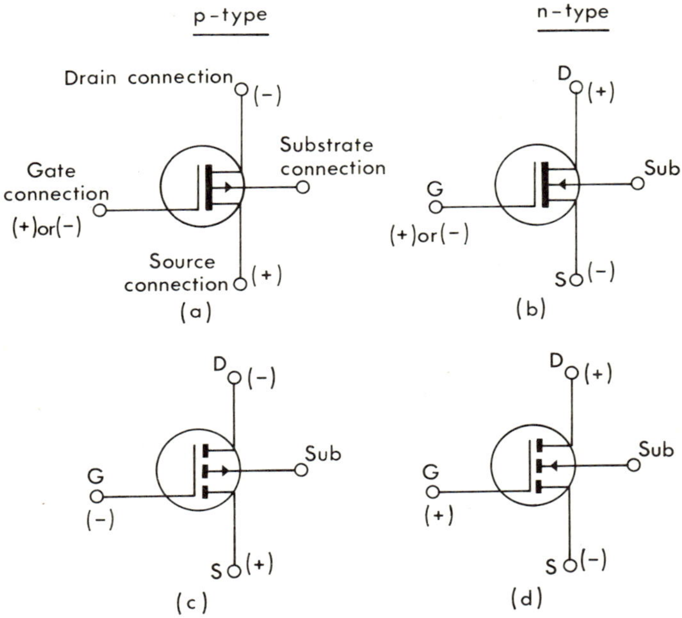

Fig. 5.4 Symbols for the four types of IGFET.
(a) p-channel depletion IGFET.
(b) n-channel depletion IGFET.
(c) p-channel enhancement IGFET.
(d) n-channel enhancement IGFET.

Fig. 5.4 also indicates the necessary polarity of the biasing voltages for correct operation of the devices as amplifiers. Note that in the case of the enhancement-type device, the polarity of the gate and drain electrodes must be alike, whereas the polarity of the gate potential for the depletion-type may be positive or negative.

5.6 Output Characteristics

Fig. 5.5 shows a suitable circuit diagram for determining the static characteristics of either the depletion or enhancement n-channel

IGFET. The output characteristics may be obtained by maintaining the gate voltage V_{GS} constant and plotting values of drain current I_D for variation of drain voltage V_{DS}. A family of such curves may be obtained by repeating the procedure for a number of discrete values of V_{GS}.

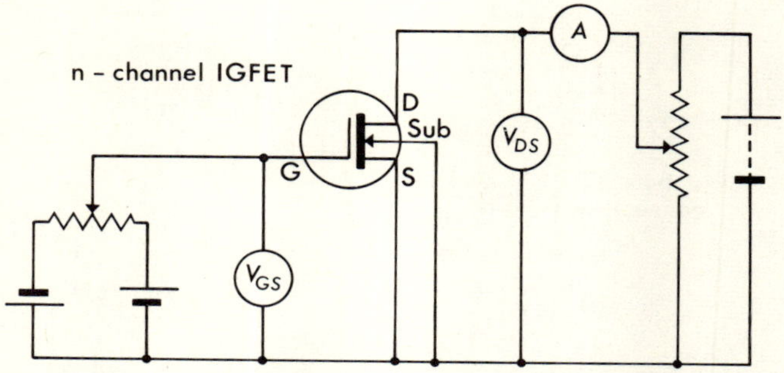

Fig. 5.5 Circuit for determination of the static characteristics of an IGFET.

Typical output characteristics are given in Fig. 5.6 for both types of IGFET. From the information given in section 5.4 on the variation of the conductivity of the channel region, it is possible to make a qualitative analysis of these curves and to appreciate the information they convey.

The actual shape of the output characteristics is very similar for both the depletion and enhancement IGFET. In fact, the only basic difference is that for the former device there exists characteristics on *both* sides of the zero gate bias characteristic.

From Fig. 5.6 it is evident that both devices exhibit two distinct regions:

(a) When $|V_{DS}|$ is less than $|V_{GS}-V_T|$. Under these conditions the IGFET acts as a low value silicon resistor and I_D increases almost linearly with V_{DS}. As $|V_{DS}|$ approaches $|V_{GS}-V_T|$ the resistance of the channel increases and the relationship between I_D and V_{DS} becomes more non-linear. This is caused by the onset of the pinch-off effect. When $|V_{DS}|$ becomes greater than $|V_{GS}-V_T|$ the saturation region is entered, and this is indicated in Fig. 5.6 by the broken line called the

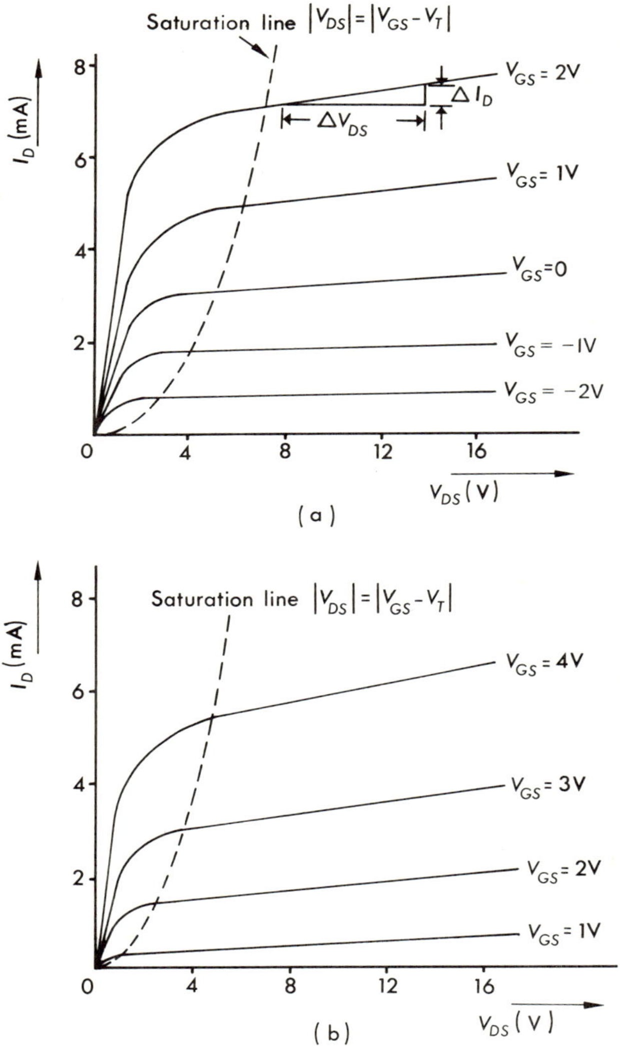

Fig. 5.6 Output characteristic curves for both types of IGFET.
(a) Depletion IGFET.
(b) Enhancement IGFET.

saturation line. This line is a plot of the points where $\mathbf{V}_{DS} = \mathbf{V}_{GS} - \mathbf{V}_T$. It marks the demarcation between the saturation region and the non-saturated region.

(b) When V_{DS} is greater than $|V_{GS} - V_T|$. Under these conditions the transistor is operating in its saturated region or **normal operating region** of its characteristic. Due to the pinch-off effect, I_D becomes almost, but not quite, independent of V_{DS}. The reason for this slight dependence of I_D on V_{DS} during saturation is primarily due to the longitudinal electric field which increases with V_{DS}. If, however, V_{DS} exceeds a certain specified magnitude, avalanche breakdown occurs and I_D increases rapidly–a condition which must be avoided.

Thus the output characteristic may be broadly divided into two regions and, like the bipolar transistor, one of these regions, the region between the vertical axis and the saturation line, must be avoided when the device is operating as an amplifier. Note that with the bipolar transistor this region was called the saturated region due to the saturated state of the base of the transistor. With the IGFET however, saturation does not occur until V_{DS} is greater than $|V_{GS} - V_T|$. Consequently this region between the vertical axis and the saturation line is called the **forbidden or active region.** Unfortunately, with IGFETs the forbidden region is wider than the saturation region of bipolar transistors and it generally requires a drain-to-source voltage of 3–4 V to avoid it—an inherent disadvantage of the IGFET.

The slope of the working part of the output characteristic is a measure of the output admittance. The **output admittance**, symbol y_{OS}, is defined as

$$y_{os} = \frac{\Delta I_D}{\Delta V_{DS}} \qquad \text{(with } V_{GS} \text{ constant).}$$

The reciprocal of the output admittance gives the output impedance, a value which depends upon V_{DS}, V_{GS} and frequency. A typical range of values for the **output resistance** r_{OS} of the IGFET is 1–100 kΩ.

5.7 Transfer Characteristics

Typical transfer characteristics for both depletion and enhancement-type IGFETs are shown in Fig. 5.7. Both curves may be obtained by use of the circuit shown in Fig. 5.5 by simply maintaining V_{DS} constant at a particular value and plotting values of I_D for variation of V_{GS}.

This characteristic is of particular interest since it indicates the control that the gate potential or signal potential has over the output current and voltage. A measure of this control is given by the slope of the transfer characteristic and is expressed as the device's **forward transadmittance** y_{fs}. When measured at low frequency (1 kHz), the forward transadmittance becomes the **forward transconductance** g_{fs} or the **mutual conductance** g_m, i.e. at 1 kHz, $y_{fs} = g_{fs} = g_m$. This very important parameter is defined as:

$$y_{fs} = \frac{\Delta I_D}{\Delta V_{GS}} \qquad \text{(with } V_{DS} \text{ constant)}.$$

Note that it has the units of amperes/volt, i.e. admittance. A typical value would be 1000–4000 μS (micro Siemen). This may also be referred to as 1–4 mA/V.

From Fig. 5.7(a) it may be seen that a small voltage V_T must be applied between gate and the source of the enhancement device before I_D flows. As already mentioned this voltage is called the **threshold voltage** V_T and is necessary to form or enhance the channel between source and drain so that I_D may flow, i.e. V_T is said to be necessary to 'turn on' the device. Notice that in Fig. 5.7(b) a 'reverse-bias' voltage of V_P (pinch-off voltage) is necessary to block I_D or to 'turn off' the device.

The terms threshold voltage V_T and pinch-off voltage V_P are synonymous. Only the former term will be used. When considering the enhancement-type IGFET, V_T may be regarded as the necessary gate voltage to cause the drain current to flow. Conversely, when considering the depletion-type IGFET, V_T may be regarded as the necessary gate voltage to reduce the drain current to zero.

5.8 Equivalent Circuit

A simplified equivalent circuit for the IGFET operating at low frequencies may be readily determined as follows. From the output characteristic (Fig. 5.6),

$$\Delta I_D = m\,\Delta V_{DS} \quad \text{for } |V_{DS}| > |V_{GS} - V_T|$$

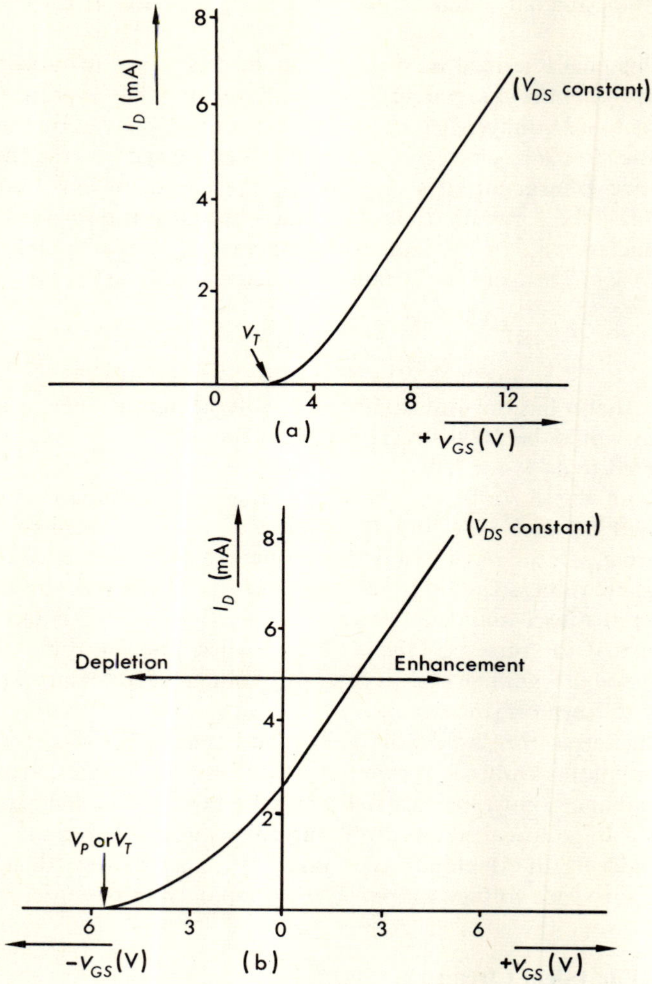

Fig. 5.7 Transfer characteristic curves for both types of IGFET.
(a) Enhancement IGFET.
(b) Depletion IGFET.

where m is the slope of the characteristic. The slope m is of course equal to the reciprocal of the output resistance r_{OS}.

If the increments of voltage ΔV_{DS} and current ΔI_D represent small signal components, then ΔV_{DS} may be written as v_{ds} and ΔI_D as i_d. Therefore

$$i_d = \frac{1}{r_{os}} v_{ds}.$$

The change in drain current caused by a change in gate voltage is a function of the transconductance of the device; in other words,

$$i_d = g_m v_{gs}.$$

Thus the total change in drain current, when both the drain and gate voltage change, is the sum of the two instantaneous changes shown above.

$$i_d = g_m v_{gs} + \frac{1}{r_{os}} v_{ds}.$$

This equation implies that the output circuit of the IGFET consists of a current generator $g_m v_{gs}$ of internal resistance r_{os} feeding an output current i_d, as a result of a change of input signal v_{gs}. Since at low frequencies the input circuit of the IGFET consists of an almost infinite, resistance the IGFET may be represented by the equivalent circuit shown in Fig. 5.8.

By taking into account the large finite input resistance and the parasitic capacitances, the IGFET may be represented by an equivalent

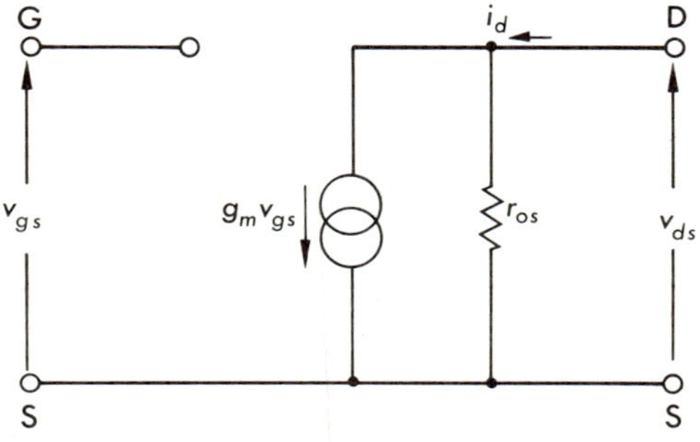

Fig. 5.8 Simplified low frequency equivalent circuit of the IGFET.

circuit which is very similar to the equivalent circuit shown for the junction FET in Fig. 4.10. However, a more representative version,

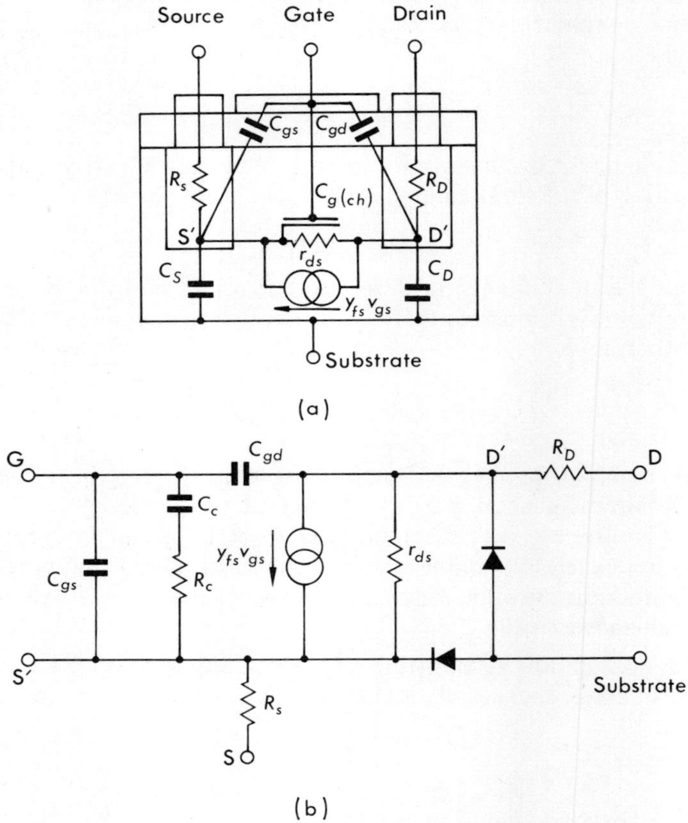

Fig. 5.9 A more representative equivalent circuit of an IGFET.
(a) Equivalent circuit superimposed on structure.
(b) More usual form of equivalent circuit.

albeit still simplified, is shown in Fig. 5.9(a). It may be seen that this circuit is derived directly from the physical structure of the IGFET.

By actually drawing the equivalent circuit superimposed on the cross-sectional view of the structure of the device, the representation of the various components making up the circuit may be more easily appreciated. The circuit is redrawn in Fig. 5.9(b) in its more conventional form. The various components represent the following:

R_S and R_D represent the bulk resistance of the source and drain plus any contact resistance.

C_{gs} and C_{gd} represent the gate-to-source and gate-to-drain capacitances.

r_{ds} represents the channel resistance which is dependent on the gate voltage.

$y_{fs} v_{gs}$ represents the output current which is a function of the gate voltage and the forward transadmittance of the device. (At low frequencies y_{fs}, g_{fs} and g_m are essentially equal).

$C_g(ch)$ represents the distributed gate-to-channel capacitance. It is shown in Fig. 5.9(b) as the series $R_c C_c$ network, where C_c represents the capacitance formed by the gate and the active region of the channel, with the silicon oxide insulating layer forming the dielectric (C_c is typically 1–5 pF) and R_c represents the series loss resistance and is of the order of 100 Ω.

C_S and C_D represent the capacitances of the pn junctions formed by the source and drain regions with substrate. These two junctions are more generally represented by two diodes connected as shown in Fig. 5.9(b).

5.9 Electrical Parameters and Ratings

Although the IGFET has been considered in a separate chapter from the junction FET, it is obvious that they are similar in many respects. This is borne out by the fact that the data supplied by manufacturers is very similar in form for all three devices.

The major electrical parameters for the IGFET are presented by manufacturers in very much the same way as that shown for the junction FET in Table 4.1. Moreover most of the major maximum ratings given in Table 4.2 are also quoted. Like the junction FET, the IGFETs are generally specified for a particular application and in consequence parameters relevant to that application are quoted.

5.10 The IGFET as an Amplifier

Although the IGFET has four terminals (including the connection to the substrate) it is generally used as a three-terminal device. In consequence it has three possible connections and as for the junction FET they are: common-source, common-gate and common-drain. Each has certain properties which gives it advantages over the others for use in particular applications. Basically the properties of the three connections are analogous to the junction FET. Fig. 5.10 shows the three configurations as simple amplifying circuits employing an n-channel, enhancement-type IGFET.

One of the major advantages that the IGFET has over the junction FET as an amplifier is that it offers a much higher input resistance—typically $10^{12}\ \Omega$. Moreover, the leakage current of the IGFET is only of the order of picaamperes (10^{-12} A) and is practically negligible. Consequently the actual input resistance of the amplifier is higher since it is not degraded by the necessity of a low-resistance return path for the leakage current.

When the IGFET is employed as an amplifier there is a definite region on its output characteristic to which operation must be confined. Such a region is shown shaded in Fig. 5.11, and is bounded by the maximum power dissipation curve, the maximum safe value of V_{DS} and the non-linear region of the characteristic where $|V_{DS}| < |V_{GS} - V_T|$.

Fig. 5.12 shows a simple low-power, common-source amplifier. In order to ensure that the d.c. operating point of the transistor lies within the permitted constant-current region of the characteristic, the gate and drain are interconnected by a high resistance, say 10 MΩ, so that virtually no current passes through it. Under these conditions the gate potential will always be approximately equal to the drain potential. In other words

$$V_{GS} = V_{DS}$$

which ensures that

$$|V_{DS}| > |V_{GS} - V_T|.$$

To avoid upsetting the bias point, the input signal v_s is coupled to the gate terminal by means of a capacitor with a suitably large capacitance, i.e. the capacitor passes the signal but acts as a block

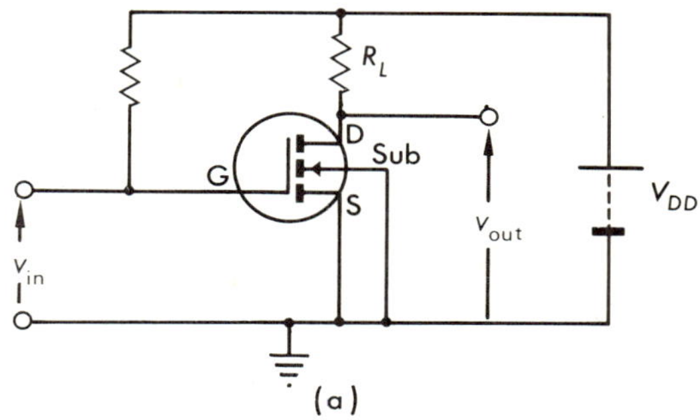

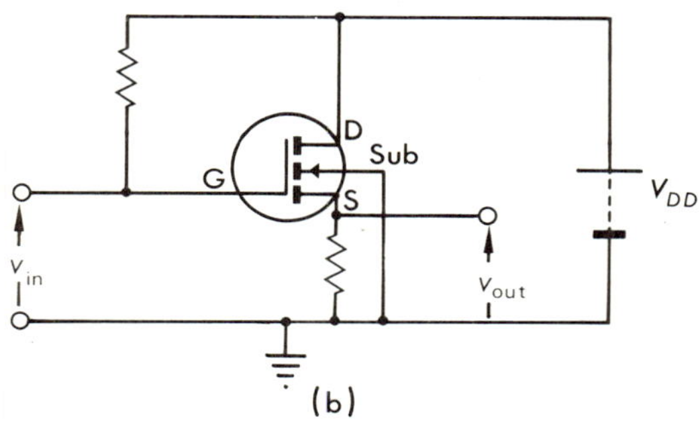

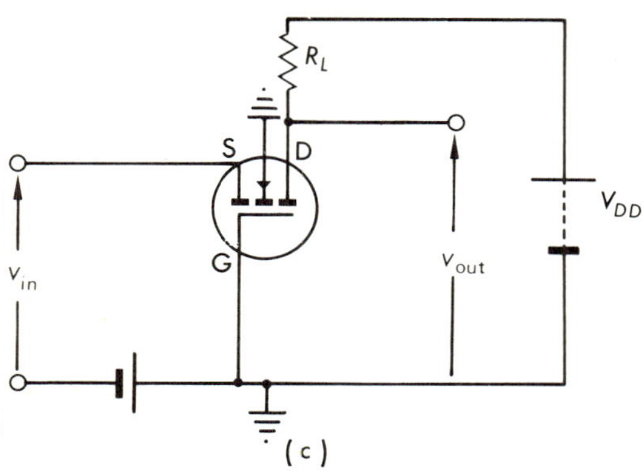

Fig. 5.10 The three possible connections for an IGFET.
(a) Common-source.
(b) Common-drain.
(c) Common-gate.

to d.c. current which would otherwise flow from the gate through the signal source resistance R_s to earth.

The value of load resistor R_L is chosen so that the resulting load line when superimposed on the output characteristic passes through the point $V_{DS} = V_{DD}$, $I_D = 0$ and cuts the $V_{DS} = V_{GS}$ curve so

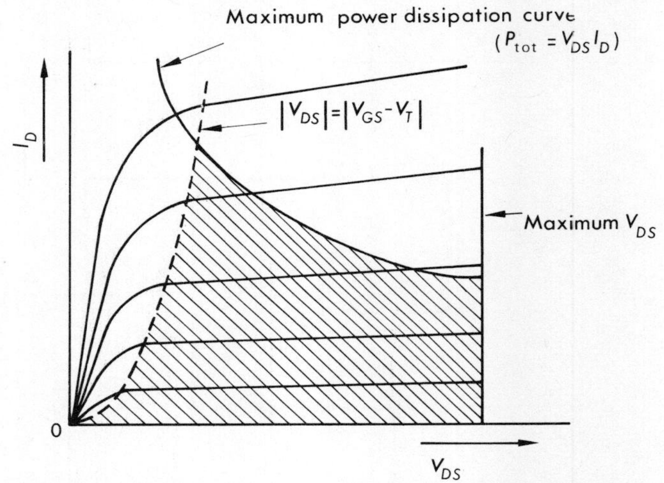

Fig. 5.11 Output characteristic of an IGFET showing permitted working area.

that the resulting quiescent current is approximately half of the maximum permissible drain current. This ensures that the d.c. operating point is well within the saturated current region of the characteristic and also that the transistor can accommodate an appreciable input signal without clipping. The value of the load resistor is given by the reciprocal of the slope of the load line.

The circuit of a corresponding simple amplifier, employing a n-channel depletion-type IGFET is shown in Fig. 5.13. The negative gate-to-source voltage is achieved by making the source positive with respect to earth by

$$V_S = I_D R_S.$$

Because the gate draws negligible current, the gate d.c. voltage is

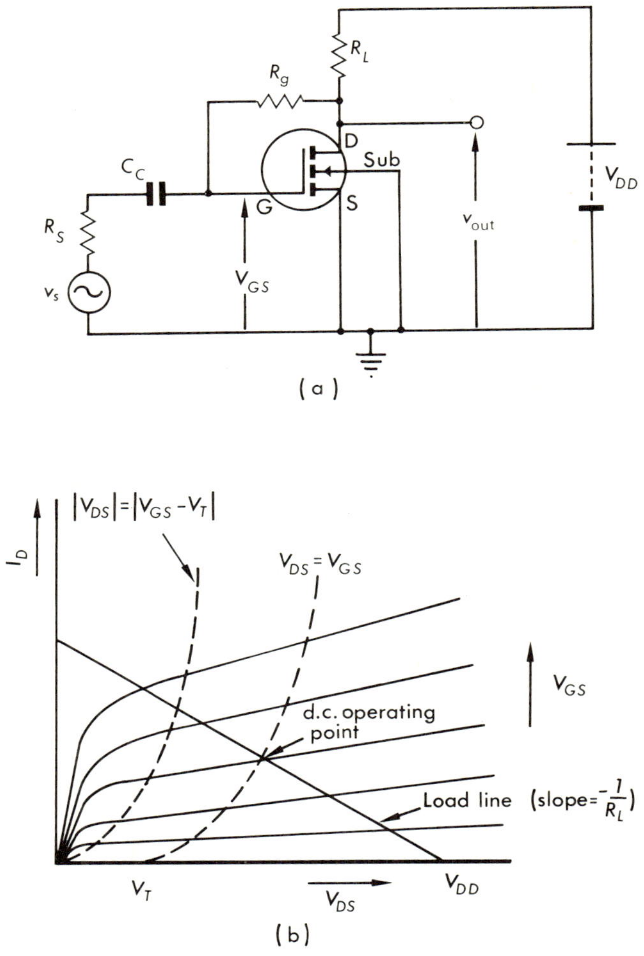

Fig. 5.12 Simple IGFET amplifier and output characteristic.
(a) Amplifier employing an enhancement-type IGFET.
(b) Output characteristic and load line.

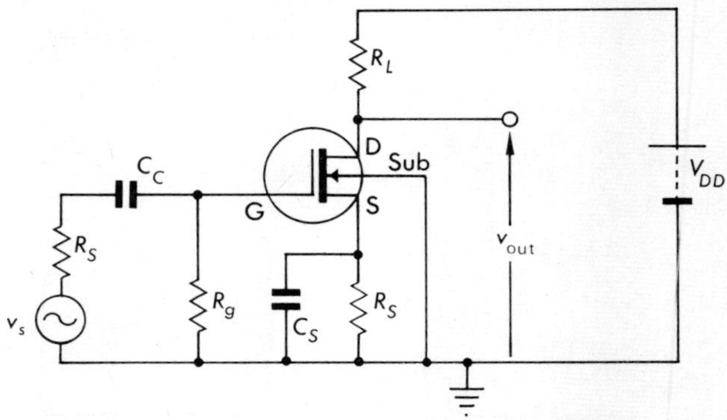

Fig. 5.13 Simple amplifier employing a depletion-type IGFET.

approximately zero (R_g is large, say 10 MΩ) hence

$$V_{GS} = I_D R_S .$$

In order to avoid negative feedback and a reduction in gain, the source resistor is bypassed by a suitably large capacitance.

Fig. 5.14 shows a more widely used IGFET amplifier on account of the increase in bias stabilization offered. The inclusion of the

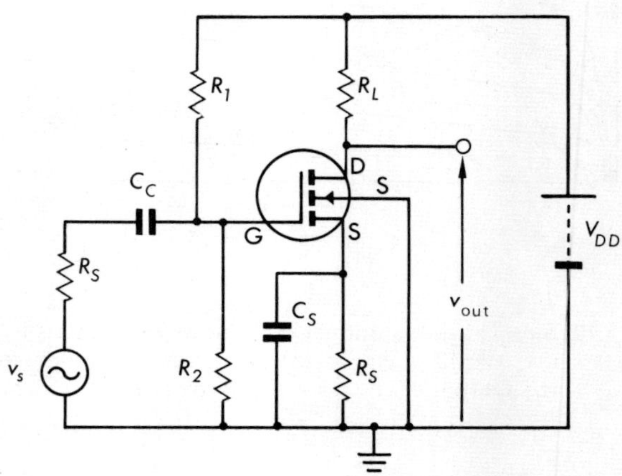

Fig. 5.14 Alternative bias arrangement for IGFET amplifier.

potential divider $R_1 R_2$ and the source resistor R_s produces d.c. feedback that stabilizes the operating point against parameter changes.

The circuit is similar to that used in bipolar transistor amplifiers, except that since the IGFET draws negligible gate current, resistors R_1 and R_2 may be considerably higher, resulting in the maintenance of a high input resistance to the amplifier.

5.11 Application of an IGFET in an Electrometer

It has already been mentioned that the gate current of an IGFET of the input current of an IGFET when used in the common-source or common-drain configuration is extraordinarily small for a semiconductor device. The very low current which does flow is due solely to the leakage in the insulating oxide layer and the leakage paths in the encapsulation of the device. It is not due to the reverse leakage current of a pn junction. With special care during the fabrication and encapsulation of the device, the gate current can be reduced to the order of 10^{-15} A. Because this gate or input current is so very small, the IGFET is used in electrometers.

The applications of electrometers lie in the measurement of minute currents and voltages. Typical examples of where such measurements are required are:

(a) Surface leakage currents in insulators and semiconductors.
(b) Ionization chambers.
(c) Thermionic valve grid currents.
(d) Currents in dimly lighted photocells.

An elementary circuit of an electrometer using an n-channel, enhancement-type IGFET is shown in Fig. 5.15. The application depicts the measurement of a very small unknown current i_x. The basic idea is to pass i_x through a very high resistor R_g (negligible shunting effect on source) and thereby cause a change of voltage in the gate circuit. This in turn causes a change in the source current of the IGFET which is detected by a sensitive galvanometer G or a centre-zero microammeter. In the interest of obtaining maximum accuracy the galvanometer current is set to zero by the variable resistor in the absence of an input current.

Since i_x is very small, say of the order of 10^{-12} A, it is essential that the amplifying device takes virtually zero current; in other words i_g must approach zero.

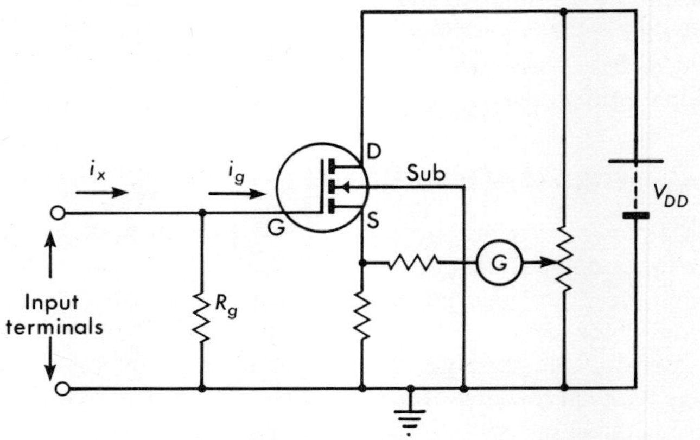

Fig. 5.15 Simple electrometer using an IGFET.

5.12 Application of an IGFET as an Electronic Switch

Although ideally an electronic switch is required to have zero impedance when in the 'on' or closed condition and infinite impedance when in the 'off' or open condition, it is possible for any device which has two distinct states to be operated as a switch. In such circuits, voltages and currents may be switched abruptly between two predetermined levels or states. These two states are generally considered as:

(a) An 'ON' state; usually represented by the device conducting heavily, implying a low impedance.
(b) An 'OFF' state; usually represented by the device being cut off, implying a high impedance.

The time taken by the device to change between the above two states is referred to as the switching time. This transition period is important since it sets a limit to how fast the switching circuit can operate. It is largely determined by such factors as the gain and frequency response of the device, the amount of drive and by the charging and discharging of the associated capacitances.

Both bipolar and unipolar transistors may be used as switches and are in fact very widely employed in digital circuit applications. They combine high reliability, high speed, low power consumption, small size, low cost and many other desirable properties. Although at present IGFET switches are slower than their bipolar counterparts, they do have the following advantages:

(a) The switching drive power is negligible. The IGFET may be made to change states by variation of its gate-to-source voltage. The current taken by the gate however is negligible, so that the drive power must also be negligible.
(b) The IGFET may be used as a normally 'on' switch (the depletion-type) or as a normally 'off' switch (the enhancement-type). No offset voltage is involved in either.
(c) The manufacture of the IGFET is potentially simpler and cheaper, and furthermore in integrated circuits the chip area occupied by an IGFET is considerably smaller than that occupied by its bipolar counterpart.

The switching characteristics of an n-channel enhancement-type IGFET are shown in Fig. 5.16. Point A occurs at low drain current and relatively high drain voltage, implying that the device is acting as a high impedance. This is representative of the switch in the 'off' or open condition. This state is obtained with zero gate-to-source voltage. If a large positive voltage is suddenly applied to the gate the operating point rapidly moves up the load line to point B. This point occurs at relatively high drain current and low drain voltage, implying a low impedance which of course is representative of the switch in the 'on' or closed position.

In both states the power dissipated in the transistor is low: in the 'on' state because of the low drain voltage and in the 'off' state because of the low drain current. Note that in making the transition from one state to the other, the operating point must pass through the active region of the IGFET, where the drain current and voltage can be appreciable. The actual energy dissipated, however, will depend upon the switching time and the switching frequency.

Frequently in integrated circuits an IGFET is employed as a high value resistor. This is because the upper limit for monolithic integrated circuit resistors is about 30 kΩ, whereas a suitably biased IGFET will act as a resistor of up to several hundred thousand

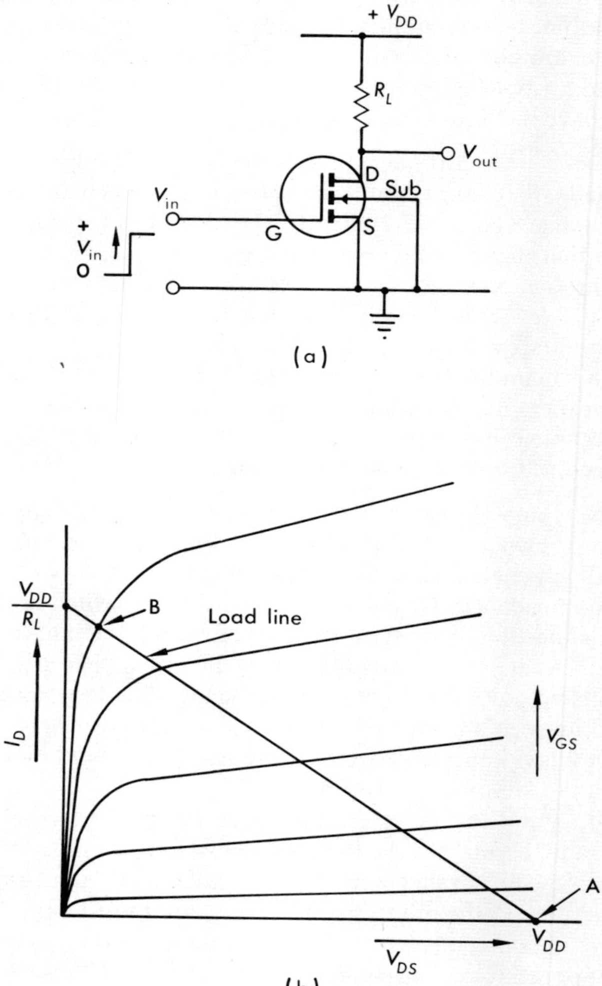

Fig. 5.16 Application of an IGFET as a switch.
(a) Circuit.
(b) Output characteristic and load line.

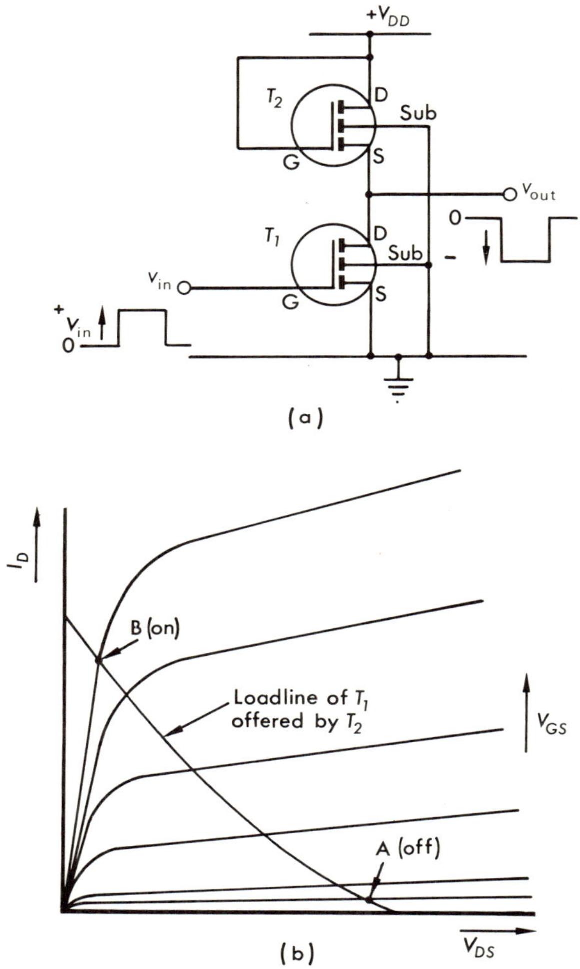

Fig. 5.17 Application of an IGFET as a resistor for a switch.
(a) Circuit.
(b) Output characteristic and load line.

ohms. Moreover the chip area occupied by an IGFET is considerably less than the area occupied by a conventional diffused integrated circuit resistor. Consequently circuit configurations such as that shown in Fig. 5.17 are common. The action of the inverter or switch is very similar to the previous circuit except that T_2 acts a non-linear resistor for T_1. To ensure that T_2 is operated in the saturated current region of its characteristic, its gate is connected directly to the drain so that

$$|V_{DS}| > |V_{GS} - V_T|.$$

Under this condition T_2 acts a resistor of 100 kΩ or more.

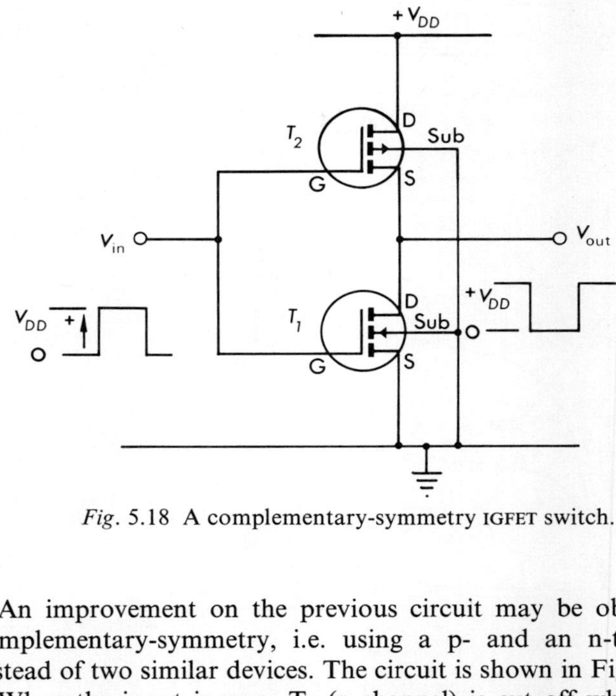

Fig. 5.18 A complementary-symmetry IGFET switch.

An improvement on the previous circuit may be obtained by complementary-symmetry, i.e. using a p- and an n-type IGFET instead of two similar devices. The circuit is shown in Fig. 5.18.

When the input is zero, T_1 (n-channel) is cut off whilst T_2 (p-channel) conducts well. The drain-to-source saturation voltage is typically less than 1 mV, therefore the output terminal is virtually connected to the supply rail and the output voltage will be very nearly V_{DD}. Conversely, when the input is made equal to V_{DD}, T_1

conducts well whilst T_2 is cut off. Thus the output terminal is virtually connected to ground and the output voltage will be approximately zero.

The advantages of this circuit are that since one or other of the two transistors is always 'on', the output impedance of the circuit is the same in both states which gives equal 'turn-on' and 'turn-off' times. Also, no current is taken from the supply except during the transition between states.

REFERENCES

CRAWFORD, R. H. *MOSFET in Circuit Design*, McGraw-Hill Book Company, 1967.

GOSLING, W. *Field Effect Transistor Applications,* Temple Press Books Ltd., 1964.

GRISWOLD, D. M. *RCA Insulated-Gate MOS Field-Effect Transistors,* Radio Corporation of America, 1966.

HOFSTEIN, S. R. and F. P. HEIMAN. 'The Silicon Insulated-gate Field Effect Transistor', *Proc. Inst. Radio Engrs.,* **51**, 1190, 1963.

RICHMAN, P. *Characteristics and Operation of MOS Field Effect Devices*, McGraw-Hill Book Company, 1967.

Chapter 6

Silicon Controlled Rectifier (Thyristor)

6.1 Introduction

The silicon controlled rectifier, usually abbreviated to SCR, is a four-layer pnpn, three-terminal device. It combines the properties of both a semiconductor diode and a transistor in such a way that it is ideally suited for a number of heavy current switching applications.

The usefulness of the SCR or thyristor may be appreciated by listing just a few of their applications:

(a) Switching currents on and off.
(b) Converting a.c. power to controlled d.c. power (rectification).
(c) Converting d.c. power to a.c. power (inversion).
(d) Controlling the a.c. power delivered to a load.
(e) Control of a.c. and d.c. machines.

The first SCR was made in the U.S.A. in 1957 nearly ten years after the invention of the transistor. Just as the transistor has been replacing thermionic vacuum valves, the SCR, since its inception, has mainly replaced the thyratron (gas-filled triode) and the low-power mercury-arc rectifier, owing to its superior characteristics, efficiency and compactness. For high-power rectification, the single-anode mercury-arc rectifier is still manufactured and will continue to be, until the thyristor can be produced at a reasonable price and with a good transient performance. Fig. 6.1 illustrates the basic structure and the symbol of the SCR.

6.2 Operation

Although the SCR is a three terminal device, a good introduction to its operation may be gained by considering only the anode and cathode connections. It was mentioned in the introduction that the

SCR combines the properties of both a diode and a transistor, so consequently our study of its operation will commence by considering a simple diode and proceed via transistor action to the actual SCR itself.

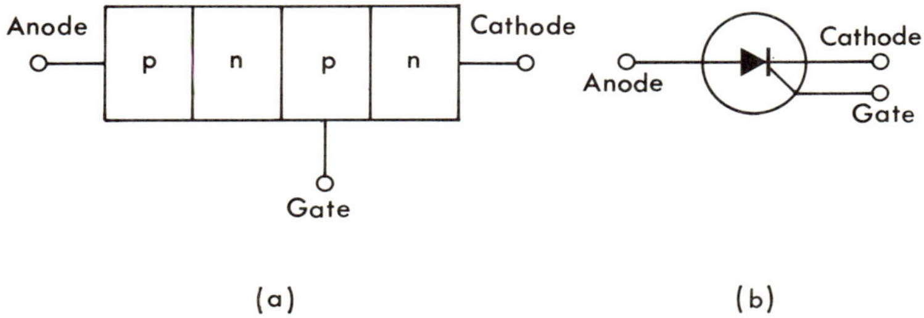

Fig. 6.1 Structure and symbol of the SCR.
(a) Structure.
(b) Symbol.

Fig. 6.2(a) shows a simple pn diode. If the positive terminal of the external voltage is connected to the n-region and the negative terminal connected to the p-region then the diode will be reverse biased. The external voltage aids the effect of the internal potential barrier and the only current which flows is the small leakage current I_{CO} due to the thermally generated minority carriers.

By adding two further layers of a p and n-type silicon to the left- and right-hand side respectively of the simple diode, the basic junction arrangement of the SCR is obtained (Fig. 6.2b). Connected to the outside p-region is the **anode electrode** whilst connected to the outside of the n-region is the **cathode electrode**.

When an external voltage is applied to the SCR so as to make the anode positive with respect to the cathode (forward bias), junctions J_1 and J_3 will become forward biased while junction J_2 will become reverse biased. Because of the presence of this single reverse-biased junction only a small current I may flow. Note that although I will be small, it will be larger than the original diode's leakage current I_{CO} due to the hole injection at J_1 and electron injection at J_3. In order to understand this carrier injection, it is helpful to consider the SCR as two interconnected pnp and npn transistors. Fig. 6.3(a)

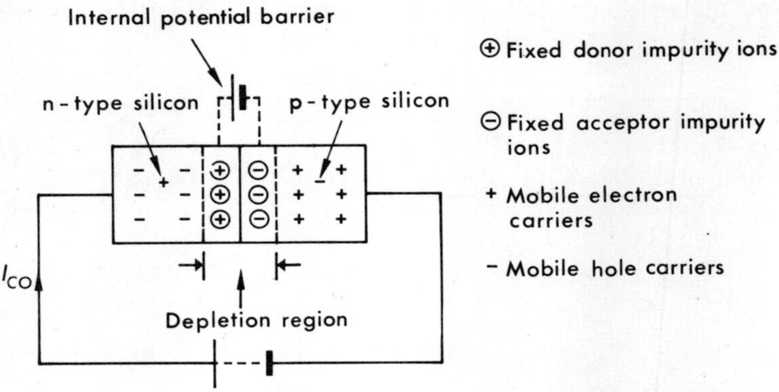

(a)

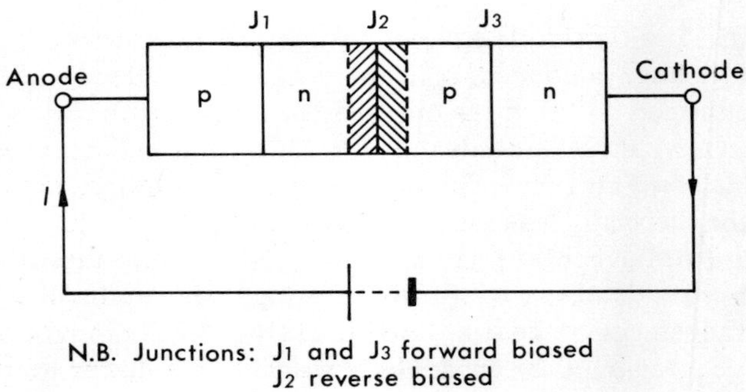

(b)

Fig. 6.2 Junction arrangement of the SCR.
(a) Simple reverse biased pn diode.
(b) p- and n-type silicon layers added to (a).

and (b) show that the SCR may be considered as two complementary transistors.

Let α_1 define the fraction of holes which are injected by the emitter and successfully reach the collector of the pnp transistor T_1. Thus, since its emitter current is I, the current which crosses junction J_2 and flows into the collector will be $\alpha_1 I$.

Let α_2 define the fraction of electrons which are injected by the emitter and successfully reach the collector of the npn transistor T_2. Again, the emitter current is I so that the current which crosses junction J_2 and flows into the collector will be $\alpha_2 I$.

The total current crossing junction J_2 consists of three components:

(a) $\alpha_1 I$, the hole current from p emitter to p collector.
(b) $\alpha_2 I$, the electron current from n emitter to n collector.
(c) I_{CO}, the leakage current of the reverse-biased junction J_2.

Since the total current must equal the external current I, we have

$$I = \alpha_1 I + \alpha_2 I + I_{CO},$$

$$I(1-\alpha_1-\alpha_2) = I_{CO},$$

$$I = \frac{I_{CO}}{1-(\alpha_1+\alpha_2)} \tag{6.1}$$

Equation (6.1) shows that the total current I flowing through an SCR depends upon the magnitude of α_1 and α_2. For a conventional bipolar transistor a typical value for α is 0·99, but this high value is only obtained by making the base region very thin, and with the fabrication of the SCR this is not possible. In order that the SCR be capable of withstanding high inverse voltages, the regions which appeared as bases in the transistor must be relatively wide, causing α_1 and α_2 to be quite low. A further important factor concerning the magnitude of α is that it is a function of the emitter current, i.e. for low values of emitter current α is also low.

Fig. 6.3(b) shows that the output of one transistor is the input of the other; this constitutes a positive feedback loop. So, unless the loop gain $(h_{FE1}+h_{FE2})$ is less than unity, an increase in the base current of one transistor initiates positive feedback, i.e. a cumulative current increase resulting in a rapid increase in the external current I.

This is borne out by equation (6.1). First consider the condition where the loop gain is low; let $\alpha_1 = \alpha_2 = 0{\cdot}25$. (The gain of the positive feedback loop is $h_{FE1}+h_{FE2}$ and since $h_{FE} = h_{FB}/(1-h_{FB})$ and $|h_{FB}| = \alpha$, $h_{FE1}+h_{FE2} = 2 \,.\, 0{\cdot}25/0{\cdot}75 = 0{\cdot}66$.) If $I_{CO} = 50\ \mu\text{A}$, then from equation (6.1)

$$I = \frac{50 \,.\, 10^{-6}}{1-0{\cdot}5}\text{A} = 100\ \mu\text{A}$$

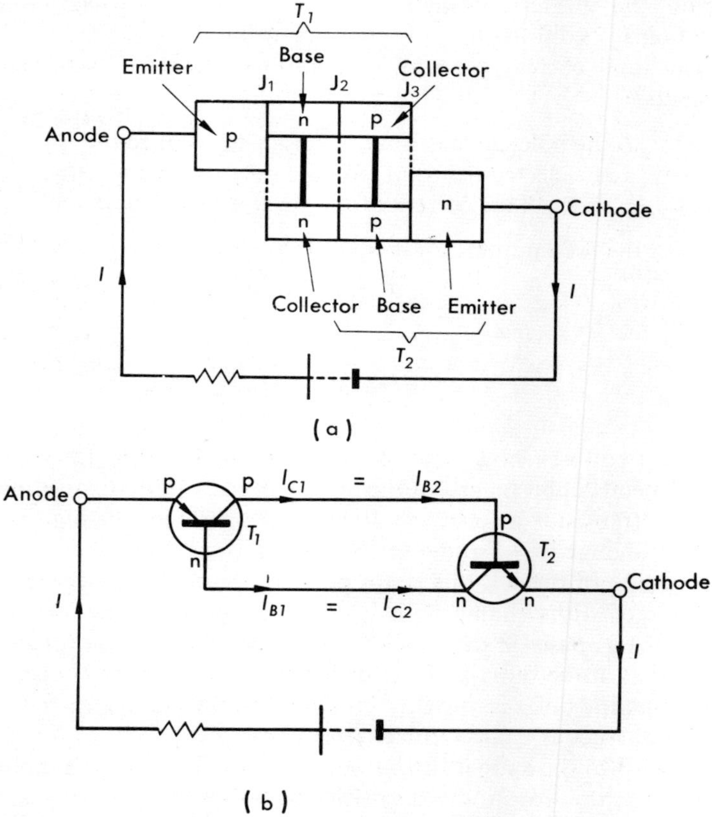

Fig. 6.3 Simulation of an SCR by a pnp and npn transistor.
(a) Four-layer device considered as two interconnected three-layer devices.
(b) Schematic representation of the two transistors.

Now consider the condition where, due to the magnitude of the applied voltage, the electric field across the reverse-biased junction is such that a few carriers are accelerated producing further current carriers on collision with the neutral atoms. Avalanche breakdown occurs resulting in the external current I increasing rapidly.

If the current I does increase because of avalanche breakdown or any other reason then, α_1 and α_2 also increase rapidly. This again is borne out by equation (6.1). Consider the condition where, owing to avalanche breakdown $(\alpha_1+\alpha_2)$ approaches unity. Then $1-(\alpha_1+\alpha_2)$ quickly approaches zero and I approaches a very large value. In practice, I must of course be limited by the circuit impedance, otherwise the device will be destroyed.

Once the SCR is conducting, the voltage across it falls (as it does with a saturated transistor). The device remains in the conducting state until the sum of α_1 and α_2 falls below unity; this may readily be obtained by reducing the current I to below a certain minimum value—known as the **holding current** I_h.

Thus, at voltages less than that required for avalanche breakdown, the current through an SCR is very small and the device is said to be **switched off.** At the critical avalanche **breakdown voltage** V_{BO}, also called the **firing voltage** V_f, the device rapidly changes state by passing a large current and in so doing becomes **switched on**. The above conditions are illustrated by Fig. 6.4. Note that if the anode is made negative with respect to the cathode (reverse-bias) junctions J_1 and J_2 are both reverse biased and the current through the device is small until the reverse avalanche breakdown voltage is reached.

6.2.1 *Gate triggering*

Although avalanche breakdown causes the SCR to switch on, an alternative method and one which is more widely used is called **gate triggering**. This is achieved by supplying additional current to the positive feedback loop by means of the third electrode called the **gate**. This gate current contributes to the current crossing junction J_2, and so increases α_2 of the component npn transistor. Thus the current gain of the npn transistor will be controlled by the anode-to-cathode current plus the injected gate current. Any increase in the base current of this transistor will be amplified and fed back regeneratively to the other transistor. Providing the gate current is large enough, the effect is cumulative. All three junctions of the SCR go into avalanche breakdown, resulting in a very rapid increase in external current and a sudden fall of potential across the SCR.

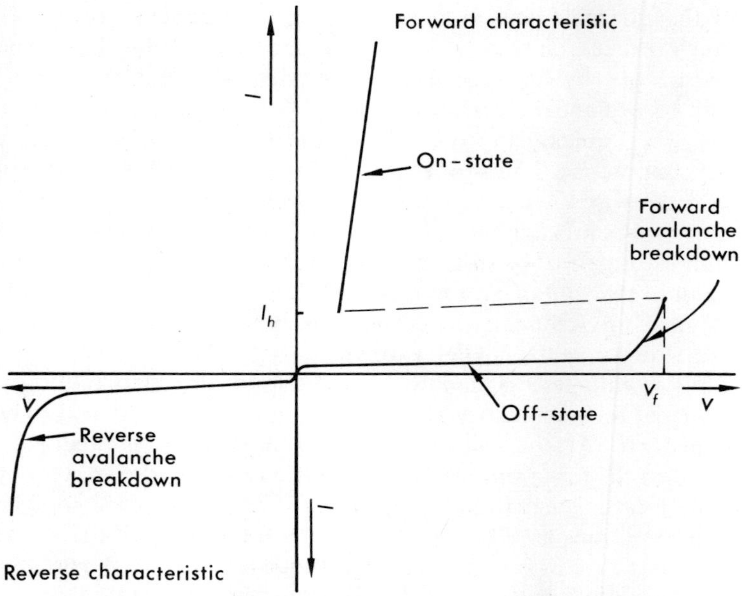

Fig. 6.4 *V/I* characteristic of the SCR (no gate triggering).

This shows that the magnitude of $(\alpha_1+\alpha_2)$ is no longer solely dependent on the anode-to-cathode current (which in its turn is dependent on the applied voltage), but that $\alpha_1+\alpha_2$ may also be controlled by the gate current. In other words, breakdown may be initiated by increasing α_2 by a suitable current pulse applied to the gate even though the applied voltage is below the original value required for avalanche breakdown. In fact, the only essential requirement for the SCR to switch on is that

$$\alpha_1+\alpha_2 = \text{unity}.$$

This may be readily achieved by increasing the applied voltage or

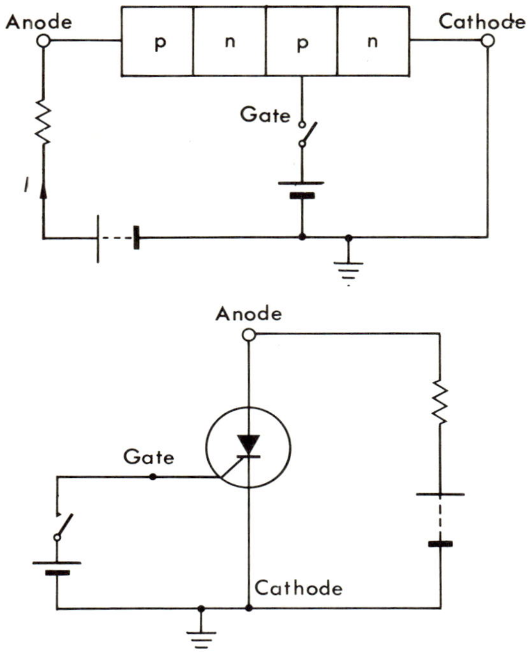

Fig. 6.5 Polarity of electrode potentials.

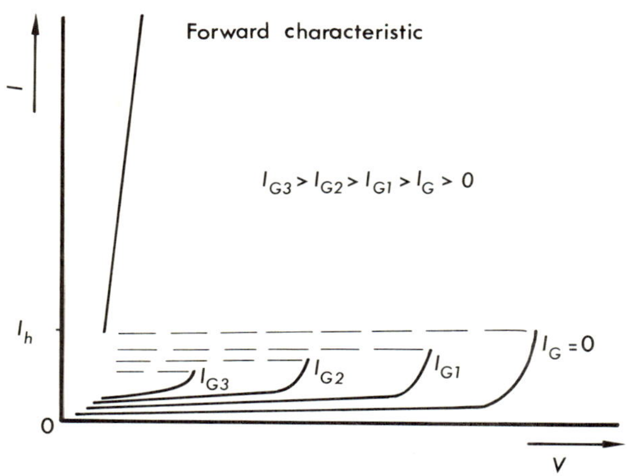

Fig. 6.6 Effect of gate current on the V/I characteristic.

increasing the gate current I_G. Fig. 6.6 illustrates a family of characteristics obtained by varying the gate current.

6.3 Typical Voltage–Current Characteristic for the SCR

Various parts of the voltage–current characteristic of the SCR have already been encountered. However, as a means of presenting a concise outline of the action of the device, a complete and typical characteristic, such as that shown in Fig. 6.7 will be briefly examined.

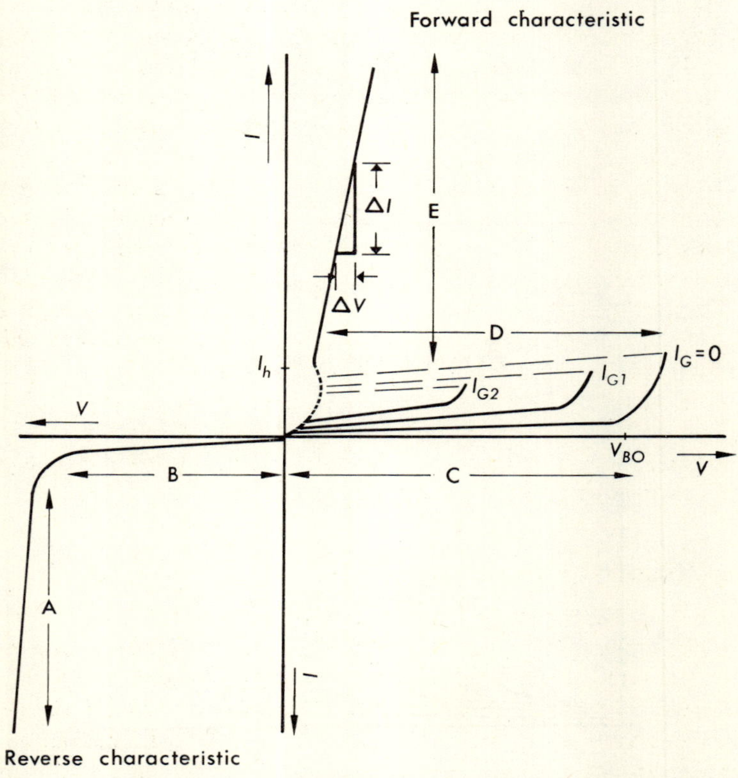

Fig. 6.7 *V*/*I* characteristic of the SCR (with gate triggering).

The characteristic may be conveniently divided into a number of important regions. When the SCR is reverse biased, i.e. the anode is at a negative potential with respect to the cathode, the **V**/I relationship is very similar to a reverse-biased diode. This is indicated by region B. If the reverse voltage is increased beyond a critical value, avalanche breakdown occurs and the reverse current increases abruptly (region A).

When the anode is made positive with respect to the cathode (forward bias) and the gate current is zero, the current through the device is small (junction J_2 is reverse biased) so that α_1 and α_2 are small and the device remains in the blocking or cut-off state—region C. If the voltage is increased to V_{BO}, avalanche breakdown occurs, $(\alpha_1+\alpha_2)$ increases rapidly towards unity, the device switch on and current abruptly increases whilst its voltage collapses from V_{BO} to a small value typically of the order of one volt. During the process of switching on, the device moves through an unstable negative resistance region D, so-called since a decrease in voltage causes an increase in current.

When in the on-state region E, the current through the SCR must be limited by external circuit resistance. The incremental resistance R_h of the device (given by $\Delta V/\Delta I$ when in the on-state is very low, rarely exceeding 10 Ω.

The device remains on for as long as the current through it exceeds I_h, the holding current, However, once the current falls below I_h the device reverts to its original off-state.

If a current I_{G1} is supplied to the gate terminal, $(\alpha_1+\alpha_2)$ is increased and the device switches over at a lower voltage than V_{BO}. An increase in gate current I_{G2} causes the device to switch on at a still lower value of forward voltage.

6.4 Parameters Controlling α

It has been seen that the factor which controls the state of the SCR is the magnitude of α_1 and α_2. In order for the device to be cut off, it is necessary for the sum of α_1 and α_2 to be less than unity. Now although this condition is readily achieved with silicon devices operating at low current, it is not obtainable with germanium devices. In other words, it is impractical to construct germanium controlled rectifiers simply because $(\alpha_1+\alpha_2)$ would always be greater than unity and the device would always switch on when forward biased. Thus, because of the importance of the magnitudes of α_1 and α_2, it is worthwhile considering the two factors which in turn control α_1 andα_2.

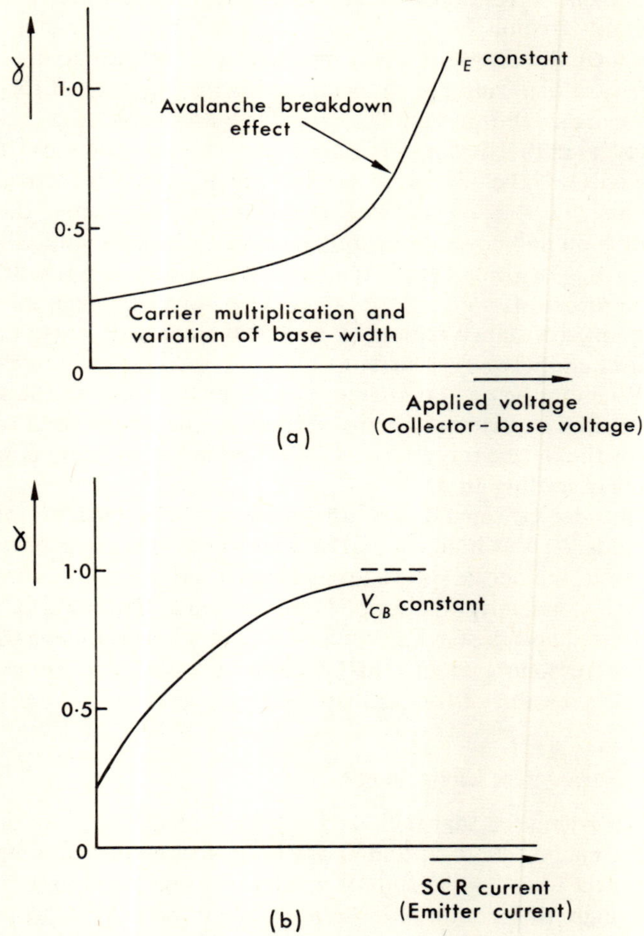

Fig. 6.8 Variation of α.
(a) Variation of α with *V*.
(b) Variation of α with *I*.

Fig. 6.8(a) illustrates how the current transfer ratio α varies with the collector-base (junction J_2) voltage. As this voltage increases, the depletion layer associated with junction J_2 increases and encroaches upon the base regions of the two component transistors. This has the effect of reducing the effective base widths, resulting in more carriers successfully reaching their respective collectors. Thus α_1 and α_2 increase.

A second and more significant contributing factor towards the increase of α_1 and α_2 with applied voltage is that as this voltage increases, some carriers which are being swept through the reverse-biased centre junction gain sufficient kinetic energy so that when they collide with an atom, they dislodge an electron from it, i.e. an electron–hole pair has been generated. These newly produced carriers may also gain sufficient energy that they too will produce further electron–hole pairs. Thus the number of carriers multiplies resulting in a greater number arriving at the collectors of the two component transistors, which implies that α_1 and α_2 are increasing with applied voltage. This phenomenon is generally referred to as **carrier multiplication.** As would be expected, if the applied voltage increases above a critical value, carrier multiplication increases to such a level that avalanche breakdown occurs and α_1 and α_2 increase abruptly.

Fig. 6.8(b) depicts how α_1 and α_2 vary with the current which is passing through the SCR. At low values of current, recombination in the relatively thick base regions of the two component transistors robs the collector current of many carriers and so α_1 and α_2 are low. As the SCR current increases, the number of carriers lost to recombination becomes a smaller fraction of the total current, consequently α_1 and α_2 increase towards unity.

Note: The current through the SCR which causes $(\alpha_1+\alpha_2)$ to approach unity so that the device starts conducting is called the **latching** or **pick-up current.**

6.5 Switching On the SCR

There are three ways in which an SCR may be switched on:

(a) By applying a positive current pulse to the gate of a forward biased SCR.

(b) By applying a forward anode to cathode voltage in excess of the avalanche breakdown voltage.
(c) By applying a forward anode to cathode voltage which has a very high rate of increase, i.e. the magnitude of the applied voltage increases rapidly with respect to time. This method is not often employed and consequently will not be discussed further.

Of these three methods, the first is by far the most widely used since it is generally convenient, rapid and effective. Frequently the required gate triggering current is less than a few per cent of the load current; moreover, when the device is switched on, the gate current may be removed without affecting the conduction of the device. Therefore a positive gate current pulse is all that is required to switch on a forward-biased SCR.

The actual magnitude and duration of the gating pulse depends upon:

(a) The magnitude of the latching current.
(b) The type of anode load.
(c) The physical size of the SCR.

In general, the requirements for the gating pulse are that its current is in the range of 100 μA to 1500 mA, its voltage is in the range of 2 to 10 V and the duration in excess of 5 μs. For pulse durations of less than 5 μs, gate current in excess of those indicated in the manufacturer's static characteristics must be used. A typical indication of how the magnitude of the gate current must increase with decrease in pulse duration below 5 μs is indicated in Fig. 6.9.

6.6 *Gate Characteristic*

In order to obtain the best performance and maximum reliability of triggering, it is obvious from previous considerations that the gate current and voltage for a given device and application are critical. Manufacturers supply a *gate characteristic* which indicates the range of gate current and gate voltage required to trigger their devices. Such a characteristic is illustrated in Fig. 6.10.

The SCR, like all other semiconductor devices, is temperature sensitive. If the temperature increases, the leakage current I_{CO} increases, which increases the forward current. When the device is in

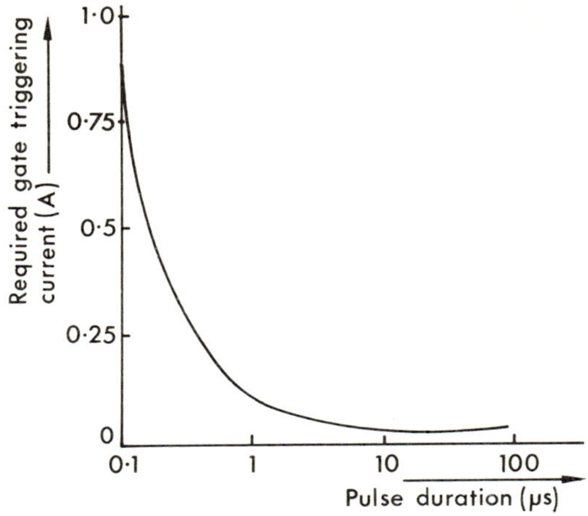

Fig. 6.9 Variation of gate current with pulse duration.

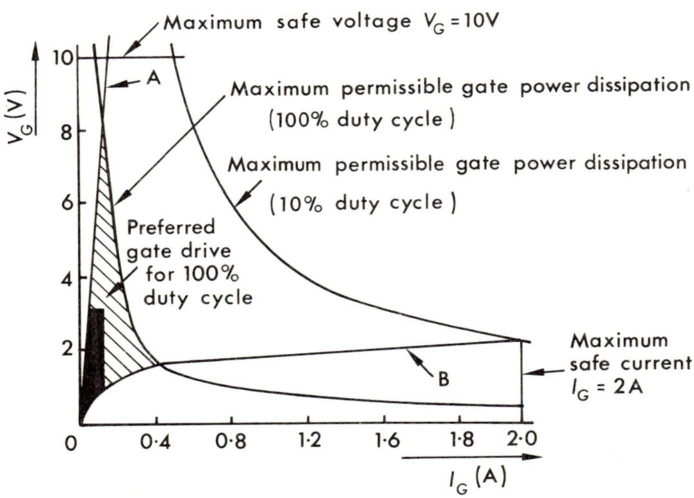

Fig. 6.10 Gate characteristic.

the off state, this increase in leakage current acts in a similar manner to the gate current in raising α_1 and α_2. Thus the trigger or gate current is a function of I_{CO}, and consequently the required triggering current must decrease with temperature increase, i.e. the trigger sensitivity increases with temperature.

The dark shaded area of Fig. 6.10 represents the range of gate current and voltage for which the SCR having this characteristic will not fire.

Curves A and B indicate the limit of gate current and gate voltage for sure firing. In other words, the area between the axes and the curves marked A and B indicates conditions under which firing is uncertain at low temperatures.

Superimposed on this gate characteristic is the maximum power dissipation curve. This is simply a plot of gate current and gate voltage whose product is equal to the average permissible gate power dissipation at the ambient temperature—usually taken as 25°C. In the case of pulse switching, where the duration of the gate pulse may be only a small fraction of the periodic time of the pulse, a second power dissipation curve is included. This curve represents the peak gate power dissipation for a pulse whose duration is one-tenth of the pulse's periodic time ($T = 1/f$), i.e. duty cycle of 10 per cent. Note that the limitation imposed by this boundary is considerably less than the limitation imposed by the curve representing continuous gate power, i.e. 100 per cent duty cycle.

To prevent the SCR from being damaged, both the peak positive voltage between gate and cathode and the peak forward current should be limited to a safe value: for the example given this is 10 V and 2 A respectively.

The area enclosed by the above boundary conditions, shown cross-hatched in Fig. 6.10, represents the preferred gate drive.

6.7 Switching Off the SCR

In order to terminate the forward conduction of the SCR it is necessary to reduce the forward current to below the holding value I_h for a given time interval so that if the forward voltage is reapplied (below V_{BO}) to the anode no current will flow. This termination may be obtained by simply reducing the forward current until it is less than I_h, or alternatively by applying a reverse voltage across the anode and cathode.

If switch-off is achieved by the first method, then the process is relatively slow owing to the saturated nature of the device. When in forward conduction the three junctions of the SCR are all forward biased, this means that the total current passing through the two

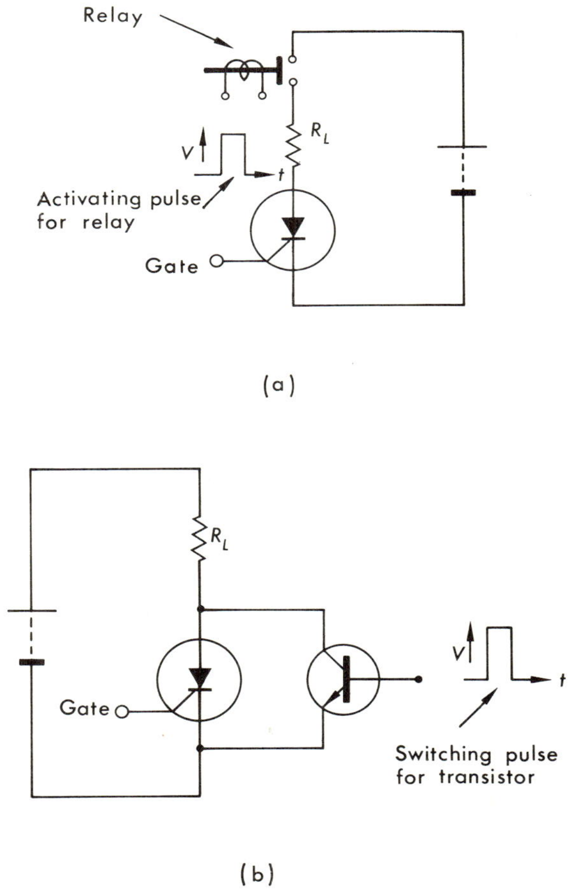

Fig. 6.11 Switching off the SCR by current interruption.
(a) Series relay.
(b) Shunt transistor.

component transistors is limited only by the impedance of the external circuit. Because the collector–base junctions of the component transistors are no longer reverse biased, there is a build up

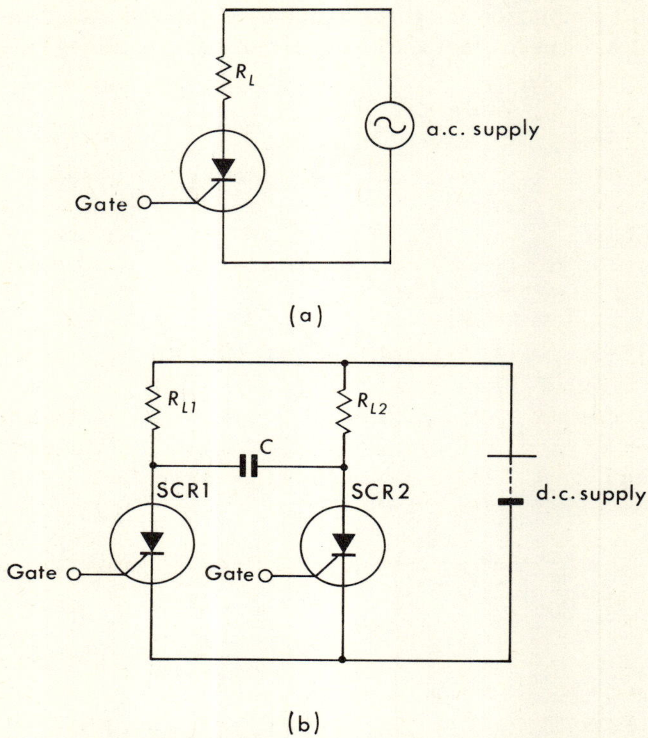

Fig. 6.12 Switching off the SCR by means of a reverse voltage.
(a) Phase switch-off.
(b) Capacitor switch-off.

or a storage of carriers within the two base regions. Consequently if the device is switched off by causing the current to fall to zero, e.g. by interrupting the applied voltage, a finite time is required for all the 'stored' carriers to disappear by recombination. A total switch-off time of anything up to 100 μs may be expected.

Fig. 6.11(a) and (b) show two possible ways in which an SCR may be switched off by interrupting its forward current. With circuit (a) it is necessary for the relay to remain open for at least 100 μs, for reasons given above. This also applies to circuit (b), that is, the transistor which is acting as a switch must remain in conduction for at least 100 μs to ensure reliable switch-off. Moreover, the collector-emitter voltage must be considerably less than the forward voltage drop across the SCR, otherwise sufficient current will pass through the SCR to maintain it in the conducting state.

If the SCR is switched off by applying a reverse voltage across the anode and cathode, then the switch-off process is much quicker, typically less than 20 μs. When the SCR is used with an a.c. supply, this a.c. source itself can be used to supply the reverse voltage, (during the negative half-cycle) necessary to switch off the device. This is depicted in Fig. 6.12(a).

If, however, the SCR operates from a d.c. supply, then special circuitry is required to produce the necessary reverse voltage. Fig. 6.12(b) illustrates such a circuit.

The action of the circuit is as follows: consider SCR1 switched on while SCR2 is off. The capacitor charges through R_{L2} and SCR1 to the supply voltage V less the small voltage drop across R_{L2}. If now a trigger pulse is applied to the gate of SCR_2 so that it switches on, the voltage across it falls to a low value. Thus the voltage across C less the voltage drop across SCR2, is applied across SCR1. The polarity is such that the anode of SCR1 is made negative with respect to the cathode, resulting in this device switching-off.

The capacitor now recharges through R_{L1} and SCR2 in the opposite sense, i.e. the right-hand plate of C becomes negatively charged with respect to the left-hand plate. Therefore when SCR1 is next triggered, SCR2 will be automatically switched off.

6.8 SCR Ratings

In general the anode–cathode ratings of an SCR are very similar to those for a silicon power junction diode. Strict attention must be paid to manufacturer's maximum current, voltage, temperature and power ratings, in order to avoid possible damage to the device.

R.M.S. forward current (I_{rms}). This is perhaps the most important rating of the SCR. This rating represents the maximum continuous r.m.s. current which may be allowed to flow between anode and cathode under certain stated conditions, such as temperature. If I_{rms} is exceeded then excessive heating within the device will result.

Average forward current (I_F). This is another basic current rating and it represents the maximum continuous d.c. current which may be allowed to flow between the anode and cathode. Its magnitude varies with temperature and conduction angle. Conduction angle refers to the fraction of the cycle over which the SCR conducts when operated with an a.c. supply.

Peak single-cycle surge current (i_{surge}). This represents the maximum non-recurrent forward current which may flow for a single cycle, without the SCR undergoing permanent damage. It is frequently referred to as the maximum fault current.

Peak forward voltage (PFV). One possible way in which the SCR may be made to switch from the off to the on state with zero gate current, is by exceeding the forward breakdown voltage V_{BO}. Although the SCR is generally gate triggered, exceeding V_{BO} is non-destructive providing breakdown occurs below a critical voltage known as the peak forward voltage.

Usually the PFV is somewhat higher than V_{BO} and its actual magnitude depends upon the device's construction and the surface properties of the semiconductor material. It also decreases with temperature decrease so that it is possible for the PFV to be lower than V_{BO} under the worst possible conditions of temperature and fabrication.

When breakdown occurs at V_{BO}, the device conducts and only a certain amount of power is dissipated in the device. If, however the peak forward voltage is exceeded, then breakdown occurs over the entire surface of the device and the resulting heating effect may cause the destruction of the device (see Fig. 6.13).

Reverse gate voltage (V_{GRM}). Just as it is important to keep within the forward characteristics, it is equally important to ensure that certain reverse characteristics are not exceeded.

The gate voltage should never be allowed to become more negative with respect to the cathode than the specified reverse peak gate voltage V_{GRM}, otherwise a considerable reverse gate current may flow should the anode–cathode loop become open-circuit—such as during switch-off. This reverse gate current would only be limited by the external gate circuit impedance and consequently it is likely that it would destroy the gate-gathode junction. If there is

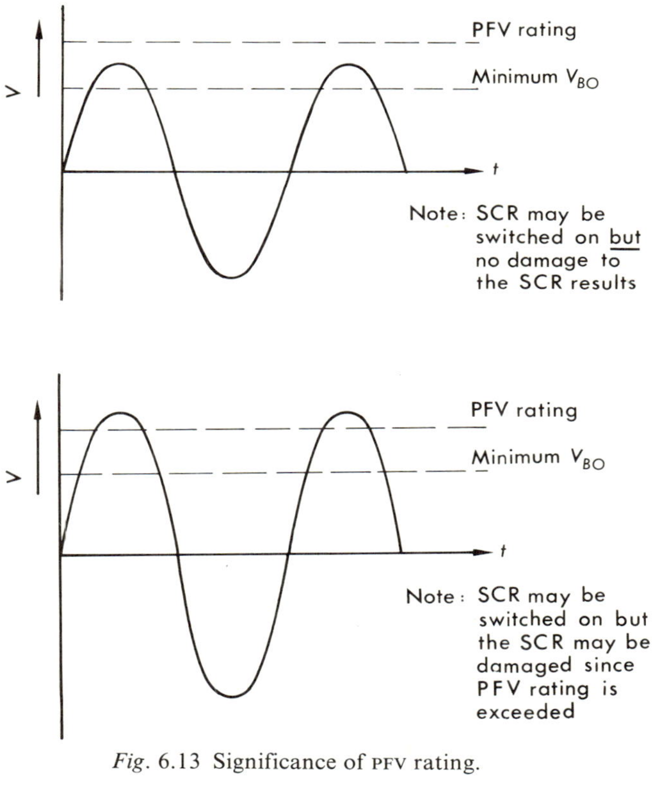

Fig. 6.13 Significance of PFV rating.

any possibility of this occurring, then a suitable diode should be connected either in series with the gate or alternatively between gate and cathode to limit any gate current to a safe value.

Peak reverse voltage (PRV). Since the SCR behaves like a conventional silicon diode when reverse biased, it follows that at a certain reverse voltage avalanche breakdown will occur, and if the resulting current is not limited, the SCR will be severely damaged. The PRV indicates the maximum allowable instantaneous value of voltage which may be applied to the anode of an SCR with the gate open-circuit.

When reverse voltage in excess of PRV are likely, a suitable diode connected in series with the SCR will avoid possibility of failure.

6.8.1 *Typical manufacturer's ratings and parameters*

Tables 6.1 and 6.2 illustrate the typical form in which manufacturers present the ratings and parameters of an SCR. This information is

TABLE 6.1 Typical maximum ratings for a low-current SCR

Peak forward voltage (PFV)	300 V
RMS forward current (I_{rms})	1·6 A
Average forward current (I_{AV})	(1)
Peak forward one-cycle surge current (i_{surge})	15 A
Peak gate power (P_{GM})	100 mW
Average gate power ($P_G(AV)$)	10 mW
Peak gate current (I_{GFM})	100 mA
Peak gate voltage (V_{GFM} and V_{GRM})	6 V
Operating temperature (T_j)	−65°C to +100°C
Storage temperature (T_{stg})	−65°C to +125°C

(1) Varies with temperature and conduction angle.

TABLE 6.2 Typical low-current SCR parameters

Parameter	Symbol	Min.	Typ.	Max.	Units
Reverse blocking current	I_{TX}	—	40	100	μA
Forward blocking current	I_{FX}	—	40	100	μA
Gate current to trigger	I_{GT}	—	5	20	μA (d.c.)
Gate supply current to trigger					
Holding current	I_h	—	0·2	1·0	mA
Switch on time	t_d+t_r	—	1·4		μs
Switch off time	t_o	—	20		μs

Note: t_d = delay time and t_r = rise time.

also supplemented by graphs which illustrate the way in which certain ratings and parameters vary under different operating conditions.

6.9 Applications of the SCR

Basically the SCR is a bistable switch, i.e. it has two stable states, one conducting and the other blocking. Although other devices may be employed as current switches the SCR by virtue of its high efficiency, speed of operation and high current and voltage ratings is generally

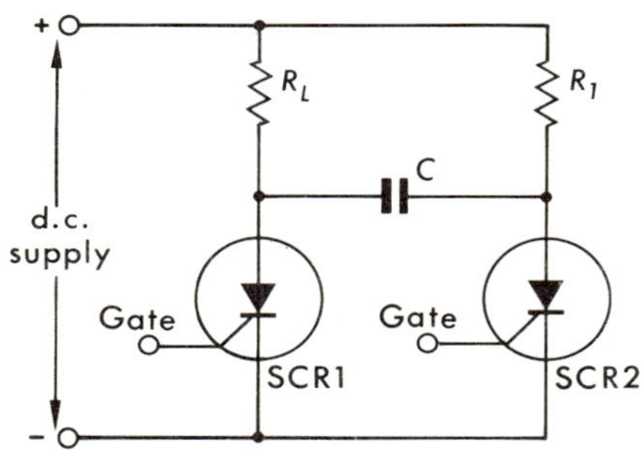

Fig. 6.14 Current switch.

superior for medium and high power operation. As an example, consider the current switching of a silicon power transistor. In order for such a device to conduct say, 5 A, a base control current of approximately 0·5 A is required. However, for the SCR the amount of current conducted is dependent on the external circuit and only a few milliamperes are necessary to switch tens of amperes. Moreover the control current maybe removed after switching. Also the wide base regions of the SCR are compatible with high voltages and these wide base regions are not possible in power transistors where thin base regions are necessary for high current gains.

6.9.1 *The* SCR *as a static current switch*

When SCR1 in Fig. 6.14 is triggered on by the application of a suitable trigger to its gate, practically the whole of the d.c. supply voltage appears across the load resistor R_L and full-load current flows. Simultaneously the capacitor C charges through R_1 and SCR1 to a voltage equal to the load voltage V_L less the small voltage drop across R_1 due to the leakage of SCR2. Thus the right-hand plate of C becomes approximately $+V_L$ with respect to the left-hand plate.

If now, it is required to switch off the load current, then SCR2 is triggered on, the voltage across it falls to a low value, so that nearly the whole of the capacitor's voltage is applied across SCR1. The polarity of this voltage is such that the anode of SCR1 is made negative with respect to its cathode, thus it becomes reverse-biased and switches off. SCR2 also switches off since the current passed by R_1 is less than the required holding current for SCR2. Hence the load current is switched off.

6.9.2 *Inversion*

Rectification is the term used to describe conversion of a.c. power to d.c. power. Inversion describes the opposite process—that of changing d.c. power into a.c. power. It is in this particular field of application that the SCR has shown itself to be so adaptable, so much so in fact that it is replacing such devices as, vibrators, thyratrons, mercury arc units, etc.

An inverter is used for d.c. to a.c. conversion, a suitable transformer may then be used to transform the voltage to a desired level then, if necessary, a rectifier may be used to convert the a.c. back to d.c., i.e. the whole system acts as a static d.c. 'transformer'.

The process of inversion is accomplished by switching and there are basically two approaches:

(a) Parallel operation, where constant current is effectively switched from one transformer winding to another.
(b) Series operation, where a voltage is switched from one resonant circuit to another.

The resulting output waveforms may be approximately square or triangular with resistive loads or sinusoidal with reactive loads. In general, the output frequency is between 25 and 2000 Hz. If externally excited, the output frequency is controlled by the excitation frequency. Alternatively if the inverter is self-excited then the output frequency will be determined by its own circuit time constant.

Fig. 6.15 shows a simple series inverter in which the SCR is used to alternatively charge and discharge a capacitor which is contained in an underdamped series resonant circuit. The load energy is supplied from the resonant circuit which also switches off the SCR.

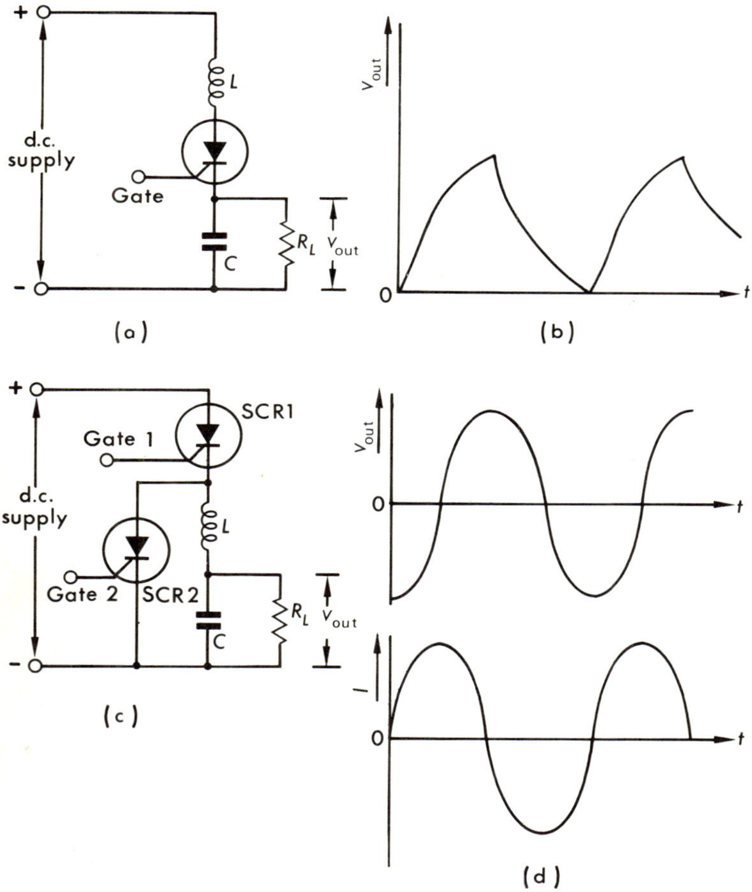

Fig. 6.15 Simple series inverter and output waveforms.
(a) Single SCR inverter.
(b) Output voltage waveform.
(c) Second SCR added to circuit (a).
(d) Improved output voltage waveform of circuit (c).

Consider the circuit shown in Fig. 6.15(a). When the SCR is triggered on, a voltage pulse is applied to the series resonant circuit which tries to go into resonance, but only one half of a sinusoidal

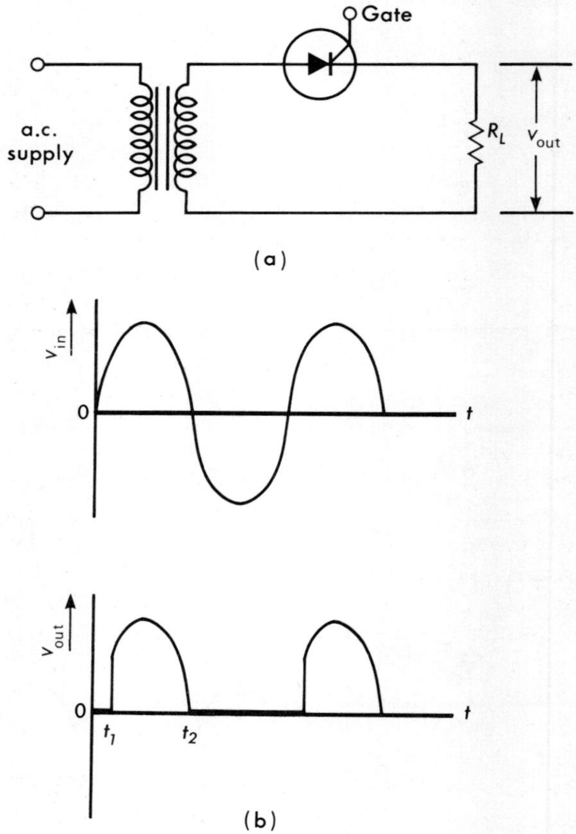

Fig. 6.16 Controlled half-wave rectification.
(a) Half-wave rectifier.
(b) Input and output voltage waveforms.

current flows because immediately the current reverses, i.e. tries to start on the negative half-cycle, the SCR switches off. The capacitor C, which is charged to approximately twice the supply voltage due to the action of the inductor L, discharges exponentially through the load resistor R_L. When the voltage has decayed to a certain level the SCR is triggered on again and the above cycle repeats itself.

Hence by the application of suitably timed trigger pulses to the gate of the SCR, the d.c. supply voltage is converted into a series of positive pulses which have a sinusoidal rise and an exponential decay. The approximate output waveform of circuit (a) is depicted in Fig. 6.15(b).

Consider now the effect of adding a second SCR, connected as shown in Fig. 6.15(c). Initially the operation is as shown above, but if SCR2 is arranged to be triggered on by G_2 at the instant the circuit reverses and switches off SCR1, then the energy stored in the capacitor C can be so discharged that a second half-cycle of sinusoidal current may flow. As soon as the current reverses a second time, SCR2 is automatically switched off and SCR1 triggered on by an external trigger to G_1. The cycle of events is repeated and the resulting alternating output voltage is shown in Fig. 6.15(d).

6.9.3 *Power control*

Before the invention of the SCR, power control was obtained chiefly through transformers and rheostats. Two types of transformer were used; those with tapped windings and Variacs. In order to attain reasonable power levels the tapped variety were large, heavy and expensive. Furthermore, since only a certain number of tappings were possible, the resulting power control was severely limited and even then only attainable in definite steps dictated by the turns ratio of the transformer. Although Variacs provided good power control, for a given power transfer they were larger, heavier and more costly.

Rheostats provided power control by introducing series resistance. This of course means that the excess power is dissipated in the rheostat in the form of heat, which in consequence results in such power control being inefficient, bulky, troublesome and limited to relatively low-power levels.

Power control by using the SCR is obtained by causing the load current to be switched on and off. For example, if only half the power available is required, then the circuit is arranged to be switched on and off, such that the total on time is only fifty per cent.

Consider the half-wave power controlled rectifier circuit shown in Fig. 6.16. On the positive half-cycle of the secondary voltage, the voltage applied to the anode of the SCR increases sinusoidally until time t_1 when the SCR is fired by a trigger pulse applied to its gate. As soon as the SCR conducts, current may flow through the secondary circuit and develop a voltage across the load resistor R_L. The SCR remains in conduction for the rest of the positive half-cycle, but as soon as the secondary voltage reverses, i.e. starts on the negative half-cycle, the SCR becomes reverse biased and switches off.

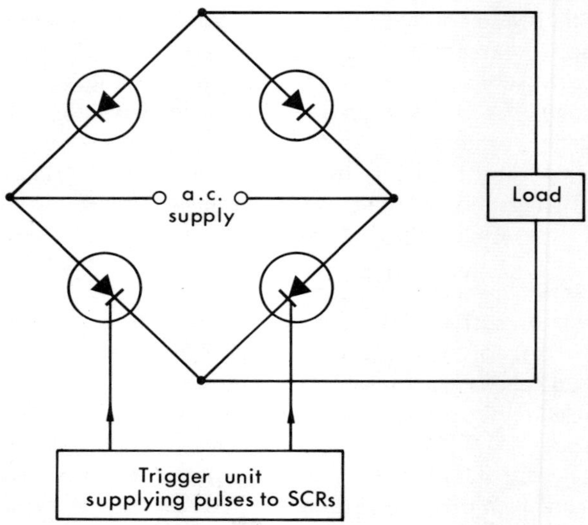

Fig. 6.17 Controlled bridge rectifier.

Thus the current through the load consists of a number of unidirectional pulses, the duration of which may be controlled by the timing of the trigger pulse applied to the gate of the SCR. Note that the smaller the duration of the conduction period $(t_2 - t_1)$ the smaller the power supplied to the load. Hence not only is the SCR rectifying, but it is also controlling the period of rectification, which in turn controls the average power supplied to the load.

Since half-wave operation is rarely employed due to its inefficiency and uneven loading effects, a more practical bridge circuit is shown in Fig. 6.17. Note that two SCRs and two diodes are employed. For most applications it is only necessary to control two of the four rectifiers as the load current always flows through two rectifiers in series. The action of the bridge rectifier has already been explained in section 2.8.4.

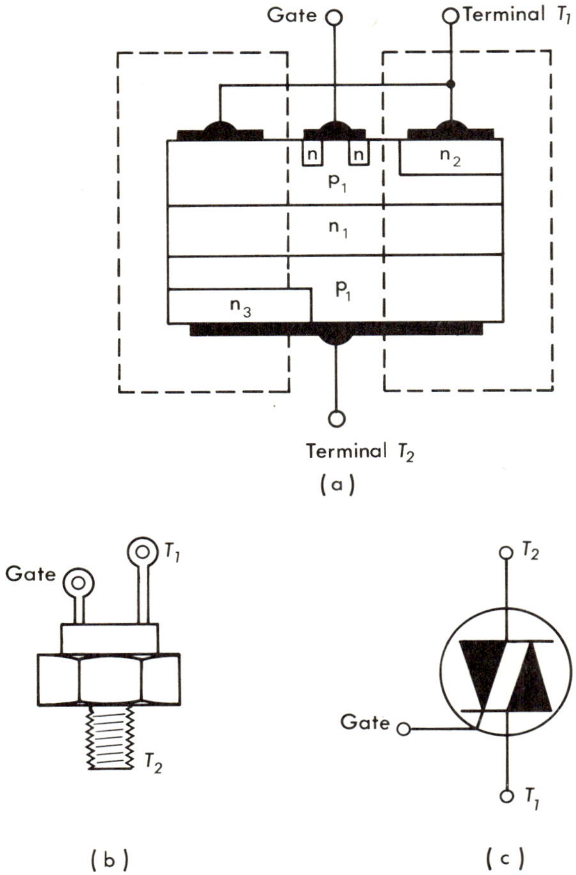

Fig. 6.18 Structure, appearance and symbol of a triac.
(a) Basic structure.
(b) Physical appearance.
(c) Symbol.

6.10 Triacs

Before concluding this chapter, it is worth considering a development of the SCR known as the **triac.** Basically this device may be thought of as a bilateral or bidirectional SCR, i.e. it has a forward *and* reverse characteristic identical to the forward characteristic of the conventional SCR. Because of this useful property, the triac has the advantage that it can block voltages of either polarity and conduct current in either direction, rather like two conventional SCRs in inverse-parallel connection.

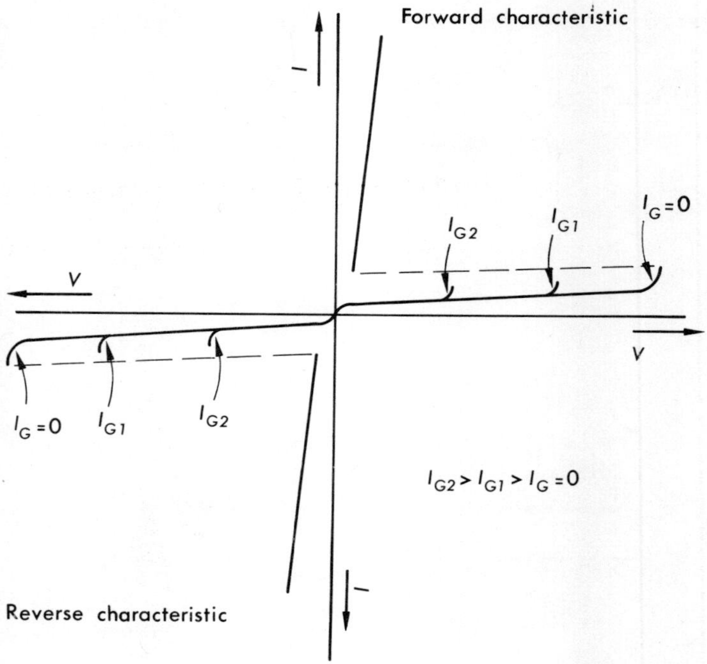

Fig. 6.19 *V/I* characteristic of a triac.

Triacs are, in general, intended for the phase control of a.c. loads in applications such as light dimming and the control of induction and universal motors.

Fig. 6.18 shows the basic structure, physical appearance and symbol of a typical triac. Fundamentally the structure is a pnpn arrangement and may be considered to consist of two conventional SCRs fabricated in inverse-parallel connection. Therefore when terminal T_2 is positive with respect to terminal T_1, a positive gate current will trigger the SCR consisting of $p_1 n_1 p_2 n_2$. The resulting characteristic will be identical to the forward characteristic of a conventional SCR. When terminal T_2 is negative with respect to terminal T_1, a negative gate current will trigger the SCR consisting of $p_2 n_1 p_1 n_3$ into conduction and the resulting characteristic will be identical to the forward characteristic but in the reverse direction. Figure 6.19 shows a typical triac characteristic.

REFERENCES

LYTEL, A. *ABC's of Silicon Controlled Rectifiers*, W. Foulsham & Co. Ltd., 1966.

Silicon Controlled Rectifier Designers' Handbook, Westinghouse Electric Corporation, 1964.

Silicon Controlled Rectifier Manual, 3rd Edn., General Electric Company, 1964.

Chapter 7

Integrated Circuits

7.1 Introduction

Almost since the beginning of the electronics industry, scientists and engineers have concerned themselves with producing smaller and smaller components in a continued effort to miniaturize electronic equipment. In 1948, a major step forward was achieved by the invention of the transistor by W. H. Brattain and I. Barden. Not only is the transistor considerably smaller that the thermionic valve it replaced, but also more reliable, cheaper and requires less power to operate.

In order to take full advantage of the transistor, passive elements such as resistors, capacitors and inductors were greatly reduced in physical size by using new materials and improved technology. The result was not only a saving in space, but, as with the transistor and semiconductor diode which motivated the trend, passive elements became more reliable and eventually cheaper.

Although considerable progress towards miniaturization was made by the invention of the transistor, it is now apparent that this invention was only the tip of the iceberg. In other words, the transistor was only a stage in the development of true microminiaturization which was made possible by the development of the **monolithic integrated circuit.** The word monolithic is derived from the Greek words monos meaning 'single' and lithos meaning 'stone'. Thus a monolithic integrated circuit is a complete circuit which is built into a single 'face of stone,' i.e. silicon.

The silicon monolithic integrated circuit is produced by the same processes used to fabricate individual transistors and diodes. The technology is merely extended so as to permit a complete circuit to be made within a single silicon chip, e.g. a typical integrated chip, measuring only 1·25 mm square by 0·25 mm thick, may contain up to fifty electronic components (transistors, diodes, resistors, capacitors) plus their interconnections.

The impact the integrated circuit has had on the electronics industry is not confined to miniaturization. It offers two other vitally important advantages over the older **discrete** component circuit, these are:

(a) Greater reliability.
(b) Production of large numbers of circuits at lower cost.

Naturally, reliability of electronic equipment has always been important. However, the ever increasing complexity of present day electrical equipment (in space probes, computers, etc.) where literally hundreds of thousands of circuit elements are utilized and where failure of any may cause complete breakdown of the system, means that greater emphasis is placed on reliability. Integrated circuits offer high reliability on account of three main factors:

(a) Absence of soldered connections between circuit components.
(b) Simultaneous fabrication of whole circuits by carefully controlled processes.
(c) Low power operation. In general, reliability of electronic apparatus decreases as the operating temperature increases. Thus, since integrated circuits are low power operated, the resulting temperature rise is small, giving high reliability. Furthermore, due to its intimate construction the effects of temperature variation are more uniform than in discrete assemblies and this contributes to greater reliability.

The reduction in manufacturing cost is achieved because many similar circuits may be fabricated simultaneously. The actual overall cost per circuit to the consumer, i.e. design cost plus manufacturing cost, does ultimately depend on the total demand for that particular circuit.

7.2 *Integrated Circuit Technology*

There are several important processes which contribute to the eventual production of the monolithic integrated circuit. They include:

(a) Purification and preparation of the silicon wafers.
(b) Epitaxial growth.
(c) Oxidation.
(d) Photomasking and etching.
(e) Diffusion.
(f) Metal deposition (interconnections).

7.2.1 *Purification and preparation of the silicon wafers*

The operation of a transistor depends upon the introduction of an exact quantity of impurities into the parent material to produce either p- or n-type silicon of the correct conductivity. It follows that since the integrated circuit may contain many transistors and diodes, the first process in the manufacture of such circuits is the purification of the silicon. The degree of purification needs to be very high, typically to a level of one impurity atom to every 10^{10} silicon atoms. This is generally obtained by a number of chemical processes. The first step involves heating the silicon compound with carbon in a furnace; the resulting material is then converted into a suitable form for purification by distillation and decomposition processes.

A further requirement of the material is that the structure of the purified silicon should have atoms aligned in an ordered array, i.e. the silicon must have, as far as is practically possible, a **single-crystal structure.** A procedure commonly employed to obtain this requirement is known as **crystal pulling.**

Fig. 7.1(a) shows a cross-sectional view of a typical piece of equipment used for crystal pulling.

Single-crystal ingots are grown by touching a suitably orientated seed crystal to the surface of purified liquid silicon contained in a crucible. The seed is rotated and very slowly withdrawn. Since the seed is at a lower temperature than the melt, heat flows from the melt to the seed. Some of the melt solidifies on the seed and the atoms arrange themselves so that they have the same crystal structure as the seed crystal.

By adding a carefully controlled amount of donor or acceptor type impurities to the melt, the resulting ingot may be either n- or p-type silicon of the desired resistivity.

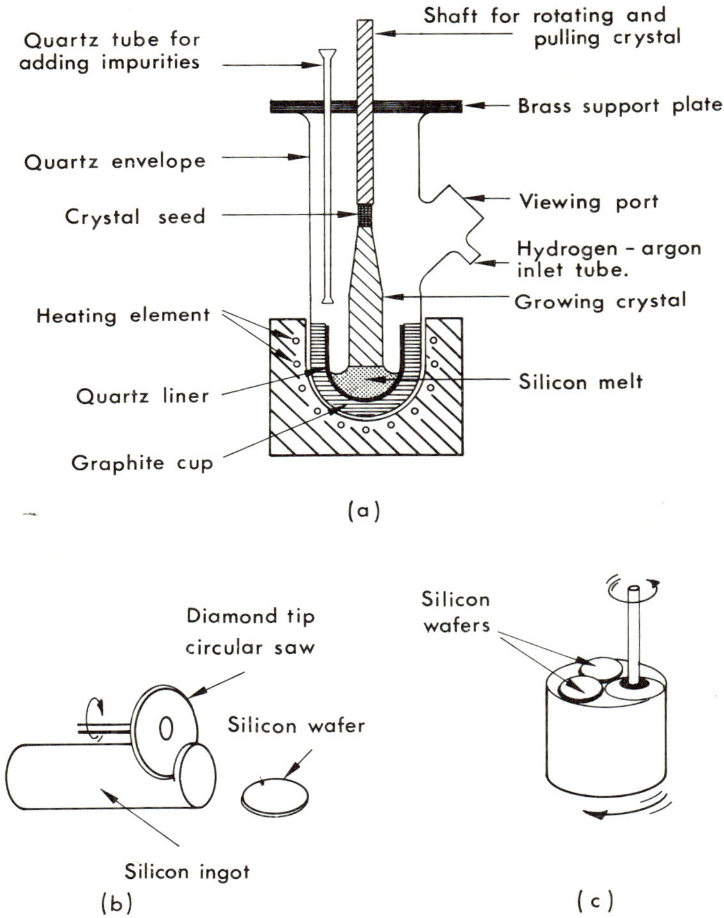

Fig. 7.1 Crystal pulling, slicing and polishing.
(a) Crystal pulling.
(b) Wafer slicing.
(c) Wafer polishing.

The ingot or rod, which is typically 15–30 cm in length and 2·5–5 cm in diameter is then ready for slicing by a diamond-tipped rotating saw (Fig. 7.1b) to give a large number of wafers approximately 0·4 mm thick. The next step in the wafer preparation is that of lapping and polishing (Fig. 7.1c) until at least one face of the wafer is suitably smooth and flat and the wafer thickness is about 0·25 mm. Finally, the wafer is chemically treated to remove any flaws and contamination of the surface caused by lapping and

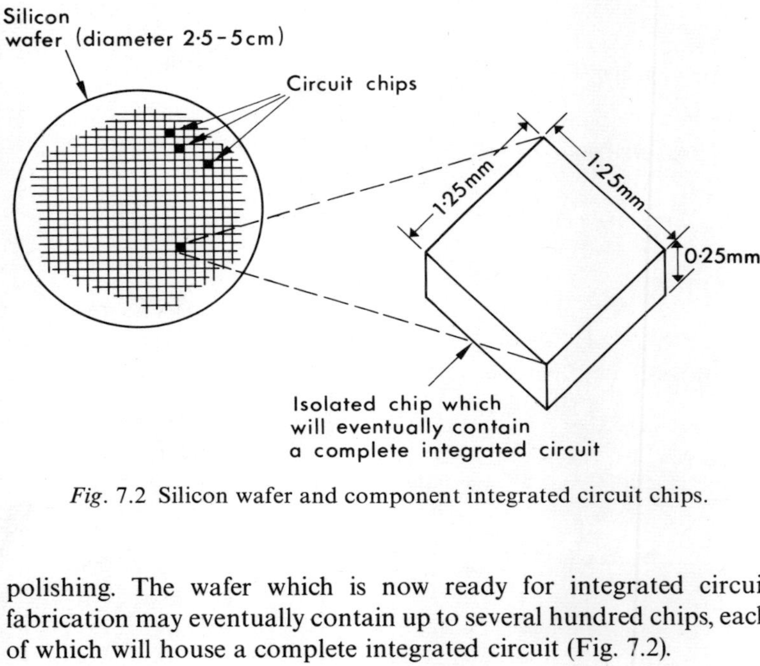

Fig. 7.2 Silicon wafer and component integrated circuit chips.

polishing. The wafer which is now ready for integrated circuit fabrication may eventually contain up to several hundred chips, each of which will house a complete integrated circuit (Fig. 7.2).

7.2.2 *Epitaxial growth*

An epitaxial growth refers to the formation of an additional single crystal layer of p- or n-type silicon on an existing silicon substrate which may be either p- or n-type. The process is used to provide a thin layer, typically 25 μm thick of low resistivity, typically 0·1 Ω-m on a substrate whose resistivity is somewhat higher (Fig. 7.3).

The epitaxial growth is obtained by exposing the substrate, i.e. wafer, to a gaseous concentration of suitably doped silicon. The grown layer has the same crystal structure and orientation as the

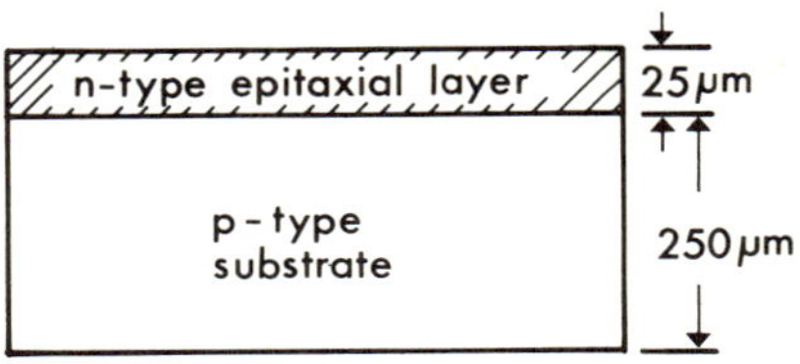

Fig. 7.3 Epitaxial layer formed on substrate.

substrate. Its resistivity is controlled very accurately by the concentration level of the dopant, whilst its thickness is determined by the time of exposure—a typical rate of growth is 2–4 μm per hour.

7.2.3 *Oxidation*

During the fabrication of an integrated circuit the working surface of the silicon wafer is frequently oxidized, i.e. a very thin film 0·5 μm thick, of silicon oxide is thermally deposited over the surface of the wafer (Fig. 7.4). The process simply involves heating the

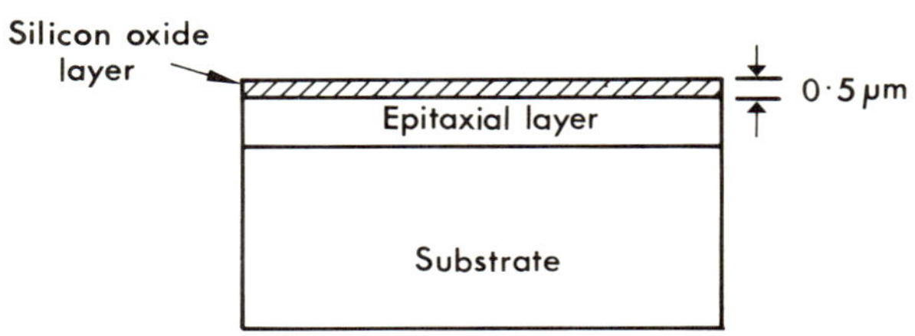

Fig. 7.4 Silicon oxide layer formed on the top surface of the wafer.

wafer to a temperature of about 1000°C and subjecting it to a flow of oxygen or steam. The thickness of the film is accurately controlled by adjustment of the temperature, moisture content and exposure time.

The reason for this process is that silicon oxide has the important property of preventing the diffusion of impurities through it. For example, if certain regions of the oxide are removed by etching and the wafer subjected to a diffusion process, the dopant will penetrate into those silicon regions only where the silicon oxide was removed, i.e. the remaining oxide serves as a mask.

7.2.4 *Photomasking and etching*

The selective removal of the silicon oxide for subsequent diffusion is achieved by means of photomasking and etching or, as it is frequently called the **photolithographic process.** The procedure is outlined in Fig. 7.5.

After the surface of the wafer has been oxidized, Fig. 7.5(a), its entire working surface is covered by a uniform film of a photosensitive emulsion, commonly called photoresist (Fig. 7.5b). Unless the photoresist is polymerized (and thereby rendered insoluble) by exposing it to ultraviolet light, it is readily removed by certain solvents. The next step involves placing on the photoresist a suitable photographic mask (Fig. 7.5c). By exposing the photoresist to ultraviolet light through the windows of the mask, the photoresist becomes polymerized (Fig. 7.5d).

The unpolymerized regions of the photoresist are easily removed leaving selected areas of the polymerized photoresist (Fig. 7.5e). This remaining emulsion is *fixed* to make it highly resistant to the etching solution of hydrofluoric acid which is used next to remove the exposed silicon oxide (Fig. 7.5f). Finally, the emulsion is removed and the wafer thoroughly washed. It is now ready for the diffusion process (Fig. 7.5g).

7.2.5 *Diffusion*

Diffusion of selected impurities into a semiconductor has been used for many years to form successive p- and n-type layers in the fabrication of high-frequency transistors. It is also a key process in the fabrication of integrated circuits.

Fig. 7.6 shows an outline of the apparatus which is used for the diffusion process. The quartz boat has a grooved platform into which the silicon wafers are placed. Underneath the platform in the bottom of the boat is the charge of the impurity element,

(a)

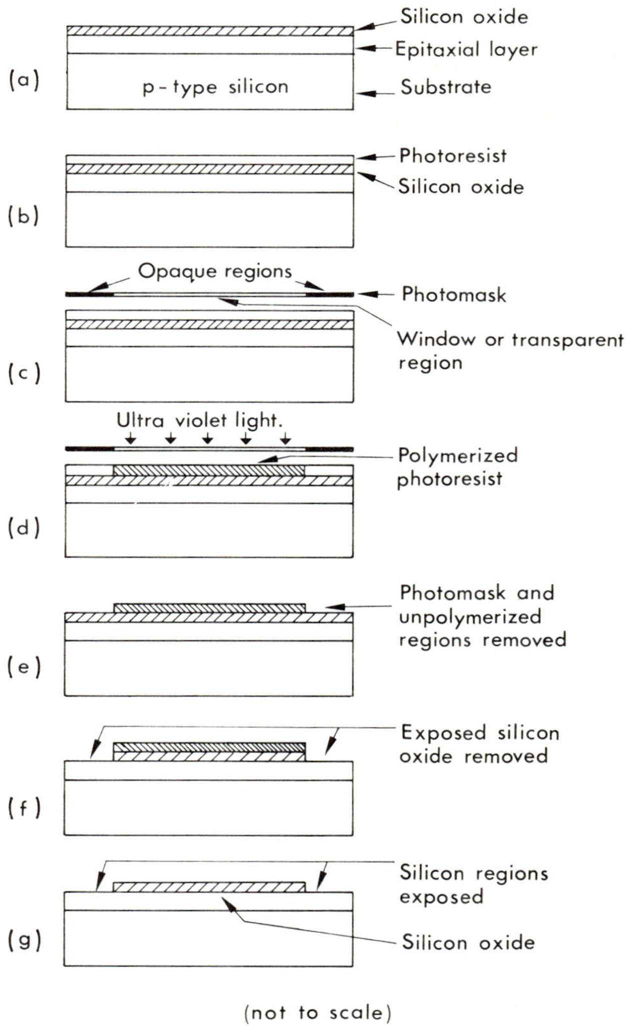

Fig. 7.5 The photolithographic process.

normally in an oxide form. The boat also has a top which is not shown.

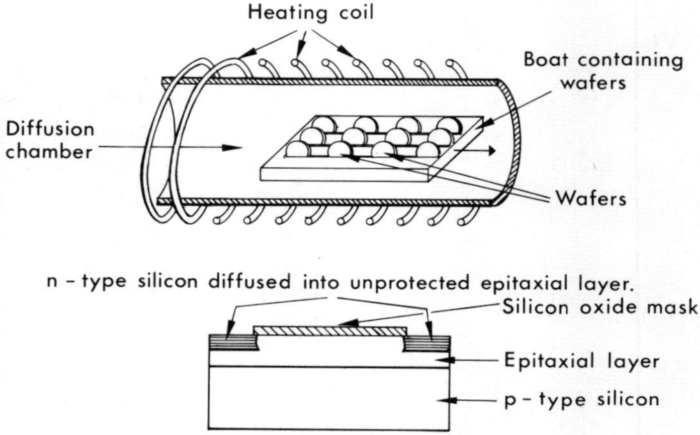

Fig. 7.6 Masked diffusion.

The boat is inserted into the diffusion furnace, where the temperature is typically 1000 to 1200°C, depending upon the dopant used. At this high temperature many of the impurity atoms will be in the gaseous state and consequently the wafers will be surrounded by a high concentration of these atoms. At the same time the atoms of the wafer will be in a highly excited state, many of them will leave their lattice and interchange with the impurity atoms, i.e. the impurities will diffuse into the wafer, their concentration being greatest at the surface. The major portion of the wafer remains unchanged.

It should be noted that the diffusion technique just explained is initially performed after the wafers have been subjected to the expitaxial growth, oxidation and photoetching processes.

7.2.6 *Metal deposition (interconnections)*

After all the circuit elements have been formed simultaneously within the wafer, it is necessary to interconnect them in a way dictated by the desired circuit. These interconnections are made by vacuum-depositing a thin, even coating of aluminium over the entire surface of the wafer. The interconnection pattern between the components is then formed by the photoresist technique. The unwanted aluminium areas are etched away leaving only the required interconnections.

7.3 Element Isolation

Because all the circuit elements—transistors, diodes, resistors and capacitors—making up a complete integrated circuit are fabricated within a common substrate which is electrically conducting, it is necessary to electrically isolate them from one another, i.e. the only interconnection between the elements must be the aluminium pattern on the surface of the chip.

There are several methods by which element isolation may be achieved. Two methods which are widely used are:

(a) Diode isolation.
(b) Oxide isolation.

7.3.1 *Diode isolation*

This method of element isolation is achieved by the process which is outlined in Fig. 7.7. The first step after the wafer has been subjected to the epitaxial (n-type layer on p-type substrate) and oxidation processes [(Fig. 7.7a)] is the selective removal of oxide by the standard photolithographic technique. The oxide mask remains only on those areas which are meant for isolation islands (Fig. 7.7b). Next, the wafer is subjected to the isolation diffusion process during which p-type impurities are diffused into the gaps in the oxide layer and permitted to penetrate through the epitaxial layer until meeting up with the p-type substrate. As may be seen from Fig. 7.7(c), n-type islands now exist on a p-type substrate.

The dimensions of these isolated islands are such that circuit elements may be formed within them. If the p-type substrate is biased positive with respect to n-type islands, good electrical insulation will exist between the component elements of the integrated circuit.

7.3.2 *Oxide isolation*

The first step in obtaining element isolation islands by this method is the formation of isolating channels (Fig. 7.8). This is achieved by using standard photoresist techniques to remove silicon oxide from predetermined boundaries and by exposing the under-lying silicon to a suitable etchant, i.e. one which will remove the silicon but not the silicon oxide. When the channels have been

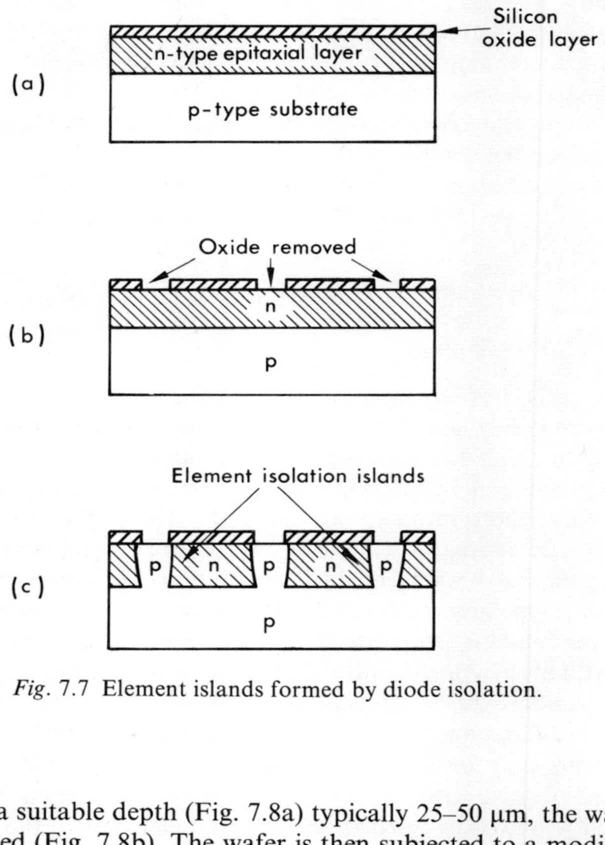

Fig. 7.7 Element islands formed by diode isolation.

etched to a suitable depth (Fig. 7.8a) typically 25–50 μm, the wafer is reoxidized (Fig. 7.8b). The wafer is then subjected to a modified epitaxial process which deposits a layer of polycrystalline silicon, typically 200–400 μm thick, on the silicon oxide (Fig. 7.8c). Finally the wafer is inverted and the original silicon is lapped down until only the single-crystal silicon islands embedded in the polycrystalline support remain (Fig. 7.8d).

7.4 Formation of Integrated Circuit Elements

Now that the fundamental processes of solid state technology have been considered, i.e. epitaxial growth, oxidation, photolithography, diffusion and aluminium metallization, it is possible to study further the procedure in the fabrication of integrated circuit elements.

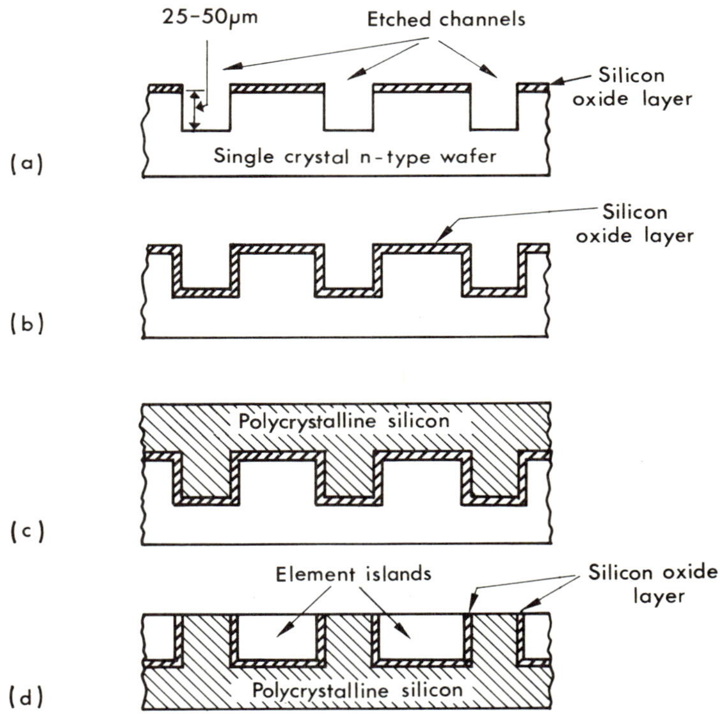

Fig. 7.8 Element islands formed by oxide isolation.

For ease of understanding, the manufacture of these elements will initially be considered individually, but it must be remembered that in practice all the elements making up a complete integrated circuit are fabricated simultaneously—in fact, hundreds of identical circuits are fabricated simultaneously.

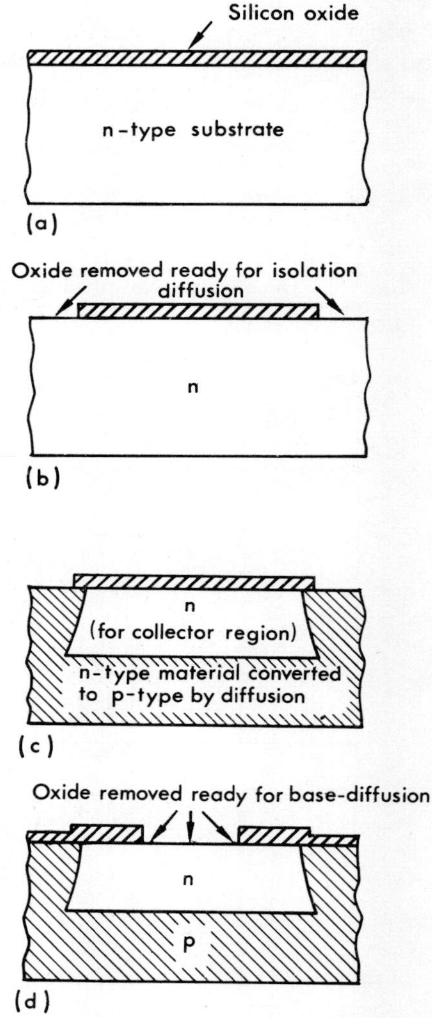

Fig. 7.9 A triple-diffused integrated circuit transistor.

(*Continued on next page*)

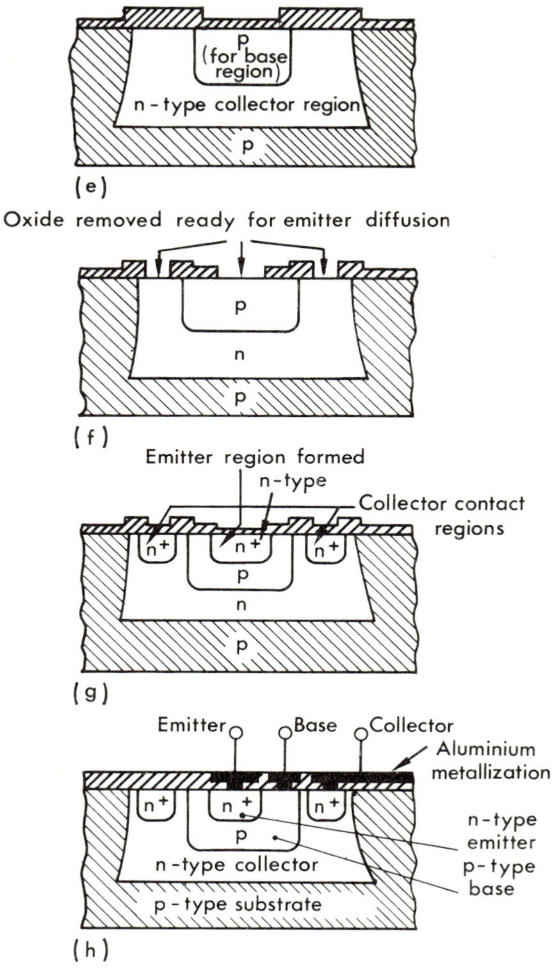

Fig. 7.9 A triple-diffused integrated circuit transistor.

7.4.1 *Transistors for integrated circuits*

Although there are several different ways in which transistors for integrated circuits can be fabricated, they are all based on the **planar process** originally developed for discrete transistors. The name planar is used because the collector, base and emitter regions are all formed through one common working surface of the wafer and the external connections all come to the same surface (planar means single surface). The planar process utilizes the various processes outlined in section 7.2.

Fig. 7.9 depicts the important steps involved in the manufacture of one widely used type of integrated circuit transistor. (Steps 1, 2 etc., do not necessarily tie up with Fig. 7.9(a), (b), etc.)

Step 1: Starting with a suitably prepared silicon wafer, a thin film of silicon oxide is thermally deposited on the working surface (top face).

Step 2: Using the standard photolithographic technique previously described, silicon oxide is removed so that the only oxide film remaining is that over the region where the transistor is to be formed.

Step 3: The wafer is subjected to heavy p-type impurity diffusion from both the top and bottom. The original n-type material is converted to p-type except for the region immediately below the silicon oxide mask. The thickness of this isolated n-type region which will eventually contain the collector, base and emitter regions is approximately half the wafer thickness. A new film of silicon oxide is deposited over the entire working surface of the wafer.

Step 4: A window in the silicon oxide is created in readiness for a second diffusion process.

Step 5: p-type impurities are diffused through the windows in the oxide until the n-type silicon immediately below the window is converted to p-type of the correct resistivity to form the base of the transistor. A further film of silicon oxide is then deposited on the working surface of the wafer.

Step 6: Further windows are etched in the silicon oxide in readiness for a third diffusion process.

Step 7: n-type impurities are diffused through the windows to form the heavily doped emitter region (n^+), plus a heavily doped (n^+) region within the collector so as to provide a good contact. Two such n^+ regions are shown, because frequently two contacts are made to the collector region. A final film of silicon oxide is laid on the working surface.

Step 8: Contact holes are etched in the oxide film and aluminium is then deposited on the wafer. This aluminium film is subsequently etched to form the separate collector, base and emitter terminals.

Compare the structure of the transistor just considered with the structure of a discrete planar transistor shown in Fig. 7.10(a). It

(a)

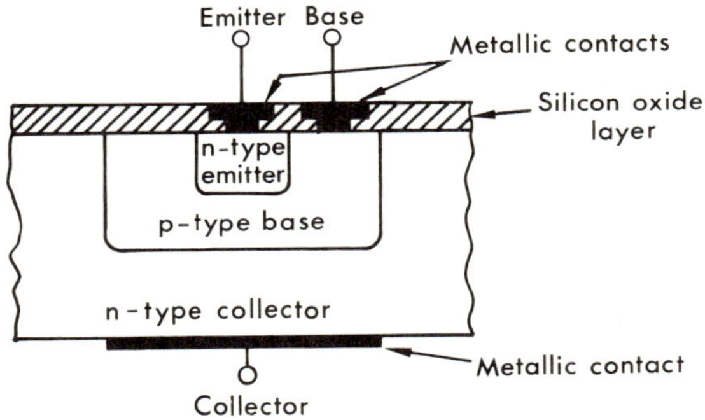

(b)

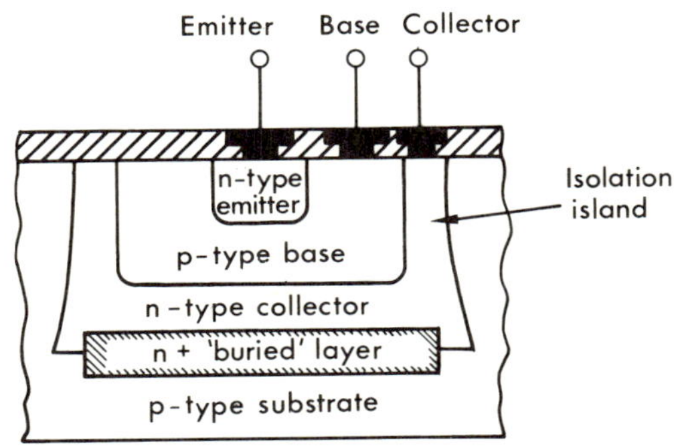

Fig. 7.10 Comparison of a discrete and an integrated circuit transistor.
(a) Discrete transistor.
(b) Integrated transistor utilizing buried n^+ layer.

will be seen that the essential difference is that with the discrete transistor the collector contact is made at the bottom of the structure, whereas it is at the top for the integrated circuit transistor. Because of the necessity for this top connection, the collector current must flow laterally along a narrow n-type region of relatively high resistivity. The result is a higher collector resistance and a higher V_{CE}(sat).

Two methods are used to overcome this high collector resistance. One approach is to use two collector contacts as suggested. Although the resistance may be reduced from say 30 Ω to 20 Ω and V_{CE} (sat) from 0·32 V to 0·20 V by using this method, the improvement is obtained at the expense of increased transistor area. A preferred solution is to diffuse (or epitaxially grow) a heavily doped (low resistivity) n^+-type silicon layer sandwiched between the p-type substrate and the n-type collector. The resulting structure is shown in Fig. 7.10(b), from which it may be seen that the collector current can now flow laterally through a low resistance region, resulting in considerable decrease in the collector resistance and V_{CE}(sat).

7.4.2 *Diodes for integrated circuits*

In the design of an integrated circuit, the resistivities (doping levels) and profiles of the various layers are generally chosen to provide optimum characteristics for the most important circuit elements—the transistors. In consequence, diodes are usually obtained by simply interconnecting the various transistor regions or by simply using a transistor pn junction. This approach is practical since the cost of a transistor in a solid-state circuit is hardly any higher than that of a diode. Several different configurations are possible. Fig. 7.11 illustrates those most widely used:

(a) Base–emitter junction; collector region shorted to base region.
(b) Base–emitter junction; collector region open.
(c) Base–collector junction; emitter region open.
(d) Base–collector junction; emitter region shorted to base region.
(e) Base–collector junction; no emitter region.
(f) Base–collector/emitter junction.

Because of the variation of resistivity and geometry of the three transistor regions, all the above configurations give diodes which have different characteristics, i.e. variation in:

(a) The forward voltage drop.
(b) The reverse breakdown voltage.
(c) The parasitic capacitances.
(d) The switching time.

The actual choice of configuration depends upon the particular circuit requirements.

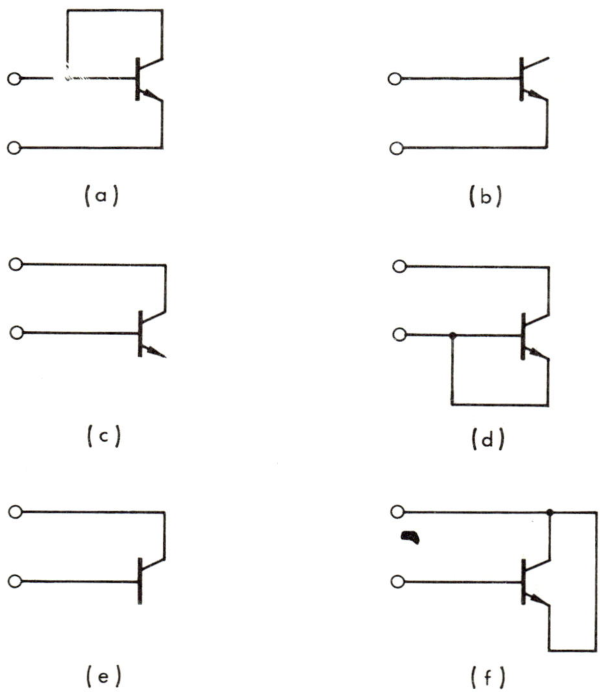

Fig. 7.11 Summary of six possible diode configurations.
(a) Base–emitter junction; collector shorted to base.
(b) Base–emitter junction; collector region open.
(c) Base–collector junction; emitter region open.
(d) Base–collector junction; emitter shorted to base.
(e) Base–collector junction; no emitter region.
(f) Base–collector junction; emitter shorted to collector.

Cross-sectional views of the structure of three diodes are depicted in Fig. 7.12, showing that the diodes are formed in exactly the same fashion as transistors. Note that the diode is formed by adjacent pn material, e.g. in Fig. 7.12(a) the diode formed has an n^+-type cathode and a p-type anode. The original n-type collector diffusion region is merely shorted to the anode so that no charge can be stored in the collector-base junction, resulting in a reduction in switching time.

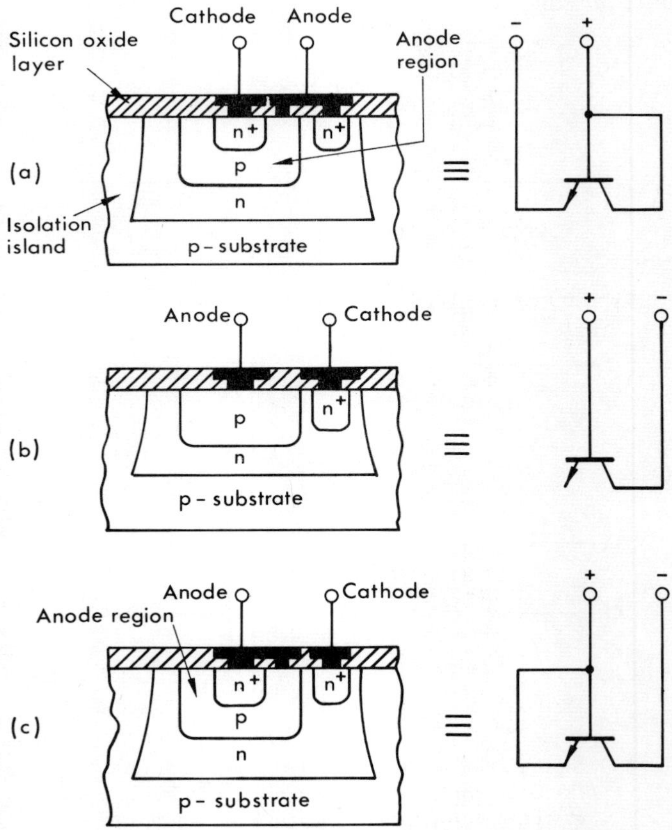

Fig. 7.12 The structure of three integrated circuit diodes.
(a) Base–emitter diode with collector shorted to base.
(b) Base–collector diode with no emitter diffusion.
(c) Base–collector diode with emitter shorted to base.

7.4.3 *Capacitors for integrated circuits*

Capacitors for integrated circuits fall into two categories:

(a) The pn junction type.
(b) The thin film or MOS type.

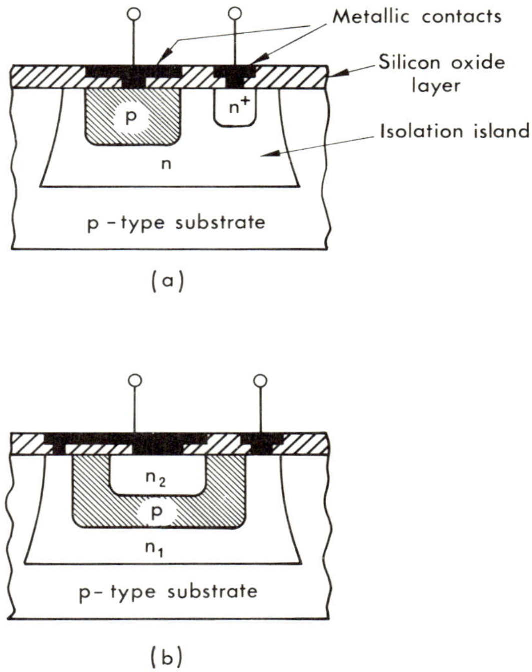

Fig. 7.13 Junction capacitors for integrated circuits.
(a) Typical junction capacitor.
(b) Junction capacitor with increased capacitance.

A pn junction capacitor uses the capacitance of a reverse-biased pn junction (section 2.5). A typical integrated-circuit pn junction capacitor is shown in Fig. 7.13(a). It is formed in the silicon chip simultaneously with transistor fabrication; thus the emitter–base junction or the collector–base junction can be utilized, e.g. for a collector–base pn capacitor, the p-type region may be diffused into the isolation n-type island at the same time as transistor base formation. The n^+-type contact region is formed into the island simultaneously with diffusion for transistor emitter regions.

The resulting capacitance associated with the pn junction depends upon:

(a) The junction area.
(b) The resistivity of the two regions utilized.
(c) The magnitude of the reverse-bias voltage.

A higher value type of junction capacitor is shown in Fig. 7.13(b). Two n-type regions n_1 and n_2 sandwich a p-type region, resulting in an increase in junction area.

From Fig. 7.13 it may be noted that an unwanted capacitor is formed by the reverse-biased n-type isolation island and the p-type substrate. Further defects of junction capacitors are their voltage dependence and the inevitable leakage current associated with reverse-biased pn junctions.

The principle of the thin-film integrated circuit capacitor is that of the parallel-plate capacitor, i.e. a capacitor is formed when two conducting plates are separated by a dielectric material. The total capacitance produced, providing the area of the plates is large in relation to the distance between them, is given by

$$C = \frac{\varepsilon_r \varepsilon_0 A}{d} \text{ farads,}$$

where ε_r is the relative permittivity of dielectric
ε_0 is the permittivity of free space ($8{\cdot}85 \times 10^{-12}$ farads/metre),
A is the area of plates in square metres,
d is the distance between plates in metres.

There are many different forms of thin film capacitors. Fig. 7.14(a) shows one type which is very widely used—it is referred to as the **metal oxide silicon capacitor** or MOS capacitor.

Plate B consists of a heavily diffused n^+ region that is diffused into the isolation island at the same time as transistor-emitter diffusion. The resistivity of this region is so low that it may be regarded as a conducting plate.

The dielectric layer is usually a film of silicon oxide of precisely controlled thickness. The high dielectric strength of silicon oxide makes it possible to have a very thin film, in the order of 0·05 μm which gives a capacitance of 600 pF/mm^2 with a safe breakdown voltage.

Plate A consists of a thin layer of aluminium that is deposited at the same time as metallization of the surface interconnections.

The advantages of this construction are its non-polarization and its simplicity of fabrication. The only additional processing step is the adjustment of the thickness of the dielectric layer.

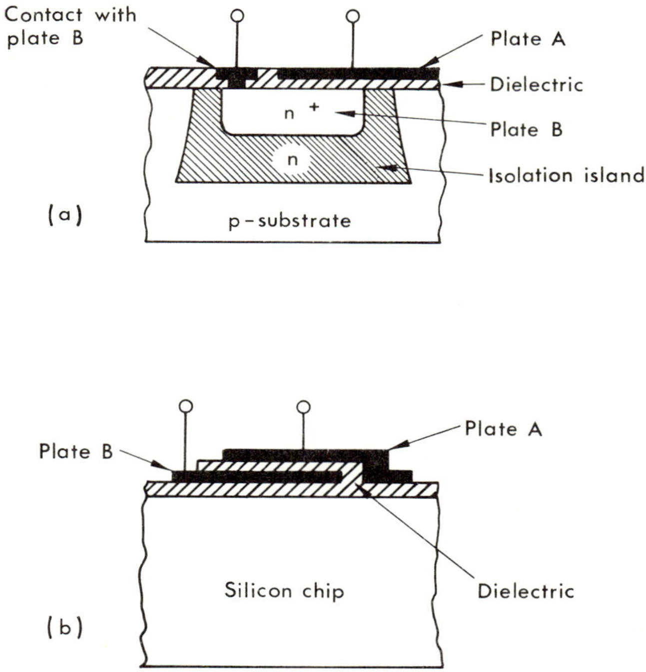

Fig. 7.14 Thin-film capacitors for integrated circuits.
(a) Metal oxide capacitor.
(b) Thin-film capacitor.

Fig. 7.14(b) shows a second type of thin-film capacitor that is fabricated entirely on top of the chip. It has an advantage over the previously considered thin-film capacitor in that it may be formed over previously diffused elements. Further advantages are that a wider choice of dielectric materials may be used and adjustment of the final capacitance may be obtained by abrasion of the top plate. These advantages are somewhat offset since the construction requires additional processing steps which adversely affect the cost and circuit yield.

7.4.4 *Resistors for integrated circuits*

Integrated circuit resistors utilize the bulk resistance of a precise volume of silicon. Depending on the required value of resistance they may be fabricated at the same time as the emitter or base regions of the transistors.

The resistance of a volume of silicon is given by

$$R = \frac{\rho l}{A}$$

where ρ is the resistivity (ohms . unit length),
l is the length,
A is the cross-sectional area.

The resistivity of the silicon, which is controlled by the impurity concentration level of the dopant, is determined by the critical requirements of the transistors. Moreover the transistors dictate the depth of penetration of the diffusion processes. In consequence, the only variables remaining to obtain a working range of resistance values are the length and width of the diffused surface. Because of this it is convenient to work in terms of a quantity known as the **sheet resistance,** symbol R_S.

The resistance of the sheet of silicon shown in Fig. 7.15(a) is given by $R_S = \rho l/wx$, where $wx = A$, the cross-sectional area. If, instead of a rectangular sheet of silicon, a square were considered, so that $w = l$, the two terms cancel each other out in the equation, resulting in

$$R_S = \frac{\rho l}{lx} = \frac{\rho}{x} \text{ ohms per square,}$$

which is known as the sheet resistance of the material.

Note that no matter how large the square may be its resistance is determined solely by the resistivity of the material and its thickness, e.g. the sheet resistance of the two squares of identical silicon shown in Fig. 7.15(b) and (c) would be the same.

The resistivity of the base region of an integrated transistor is typically 100 ohms per square. Therefore in order to obtain a resistor of say, 500 Ω, five squares are connected end-to-end. The squares may be 25 μm by 25 μm, so that the overall size of the 500 Ω diffused resistor will be 25 μm wide and 125 μm long. In cases where a large value of resistance is required, the resulting pattern formed by the relatively long length of resistor may often take on odd shapes in order to fit within a given area.

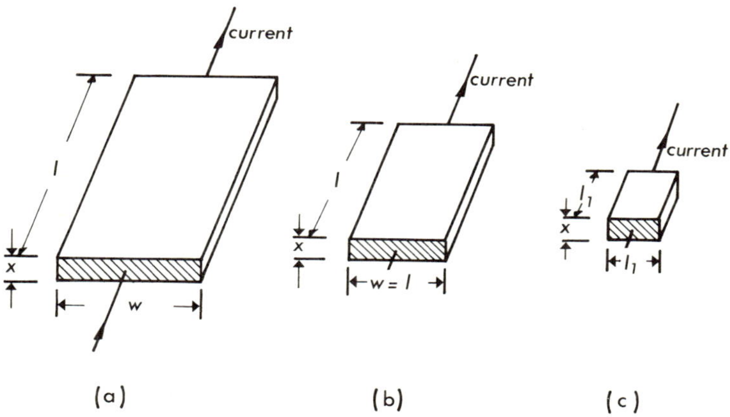

Fig. 7.15 Sheet resistance.

Most resistors are formed at the same time as the transistor base-diffusion process since the resulting resistivity of the silicon makes possible a wide range of practical resistances, e.g. 100 Ω to 30 kΩ. The emitter diffusion process produces a low resistivity region which makes possible practical resistors from 5 Ω up to 1 kΩ. The n-type epitaxial layer used for the collector region is not generally employed for resistor fabrication, the reasons being that its sheet resistance is similar to the base region and also that because adjacent to the substrate a large distributed capacitance is unavoidable.

It is difficult to produce integrated circuit resistors which are closer than ±10 per cent to the desired value. However, in circuits where resistors with tolerances better than ± 10 per cent are mandatory, it is possible to utilize two resistors formed side by side, the ratio between them being controllable to within ± 1 per cent.

The two major sources of variation in integrated resistors values are:

(a) Diffusion tolerances. It is very difficult to control precisely either the impurity concentration level of the dopant, or the resulting depth of penetration of the diffusion process. variation of these two factors account for most of the variation in resistance values.

(b) Photomasking and etching tolerances. Since the width of a resistor is typically only 25 μm, variations in the dimensions of the windows of the mask or misalignment of the mask on the wafers will result in variation in the value of the resistances produced. After the photoresist process, unwanted silicon oxide must be removed before subsequent diffusion processes. Any under- or over-etching will result in further variations in the resistance of the resulting elements.

Both a cross-sectional view and a top view of a diffused resistor are shown in Fig. 7.16. The processes involved are, of course, identical with transistor fabrication, the actual resistor layer being formed, simultaneously with the base regions, by diffusion.

Note that there is a third terminal connected to the n-type isolation island, and this is provided so that the isolation island may be connected to a suitable positive voltage to ensure that the parasitic transistor formed by the pnp structure (substrate acting as collector) has both its junctions reverse-biased and therefore rendered inoperative.

7.5 Formation of a Simple Integrated Circuit

Having examined the basic processes involved in integrated circuit technology and the individual fabrication of the circuit elements, as a summary the simultaneous fabrication of the simple circuit shown in Fig. 7.17 will be considered.

Step 1: Starting with a suitably prepared p-type wafer, whose resistivity is of the order of 0·05–0·1 Ω-m an n-type epitaxial layer of resistivity 0·002–0·005 Ω-m is grown into the substrate. This

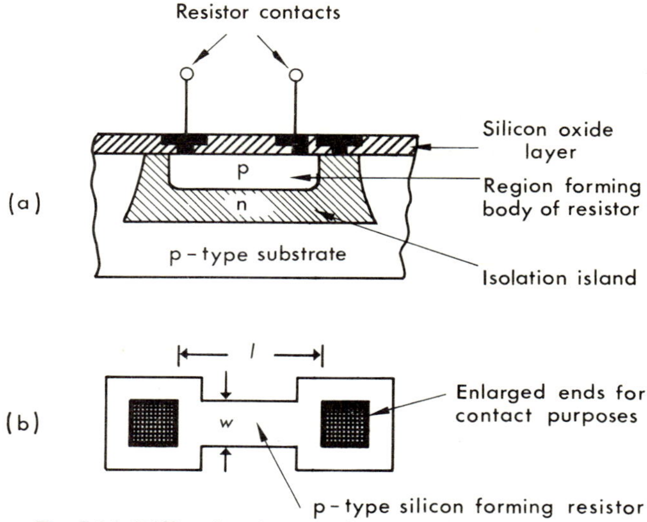

Fig. 7.16 Diffused resistor for integrated circuit.
(a) Cross-sectional view.
(b) Top view.

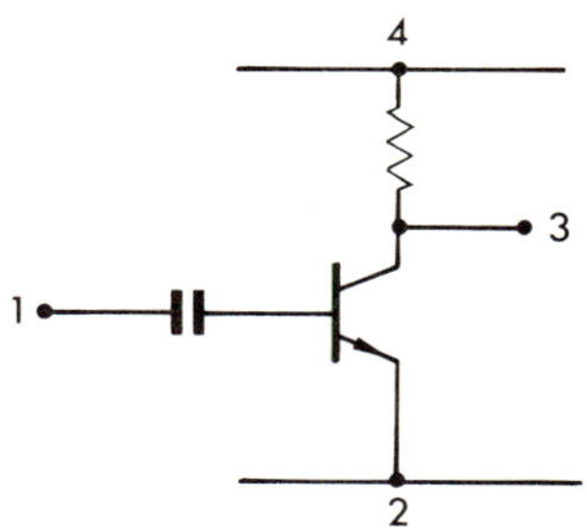

Fig. 7.17 Simple transistor stage.

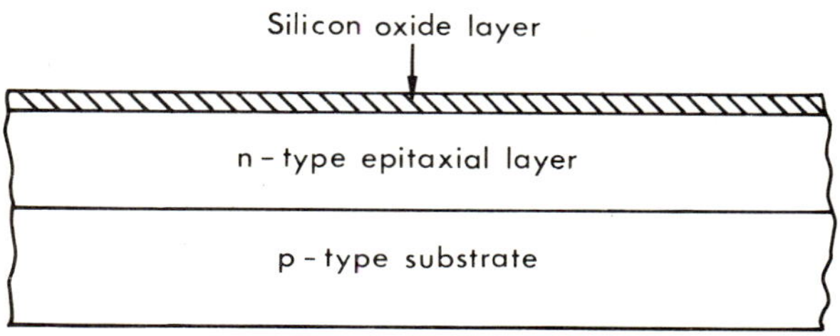

Fig. 7.18 Oxidized p-type substrate containing an n-type epitaxial layer.

layer, in which all the circuit elements are formed, is typically 25 μm thick (Fig. 7.18 is not to scale). A very thin layer of silicon oxide, approximately 0·5 μm thick, is thermally deposited over the entire working surface of the wafer.

Step 2: By standard photoresist and etching techniques, suitably situated and dimensioned windows are produced in the oxide layer and the wafer subjected to p-type isolation diffusion, i.e. isolation islands are provided for each circuit element (Fig. 7.19). The working surface is again re-oxidized.

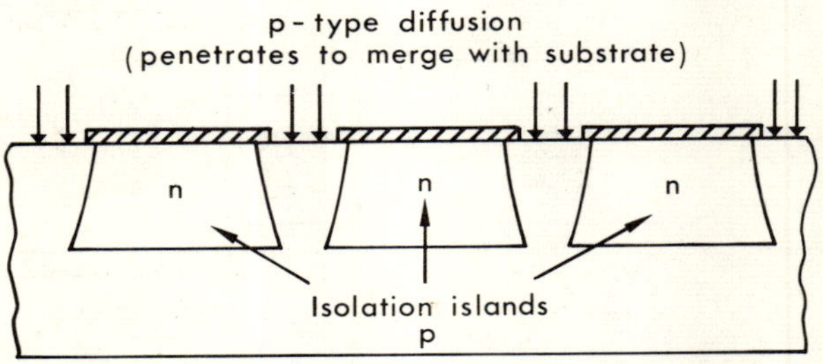

Fig. 7.19 Formation of isolation islands.

Step 3: The wafer is subjected to a second diffusion process, generally referred to as transistor-base diffusion, since it is the requirements of the base which determine the impurity concentration and depth of penetration. The p-type impurities are diffused through precisely-situated and dimensioned windows to form the base of the transistor, one half of the capacitor and the body of the resistor, as shown in Fig. 7.20. The wafer is then re-oxidized.

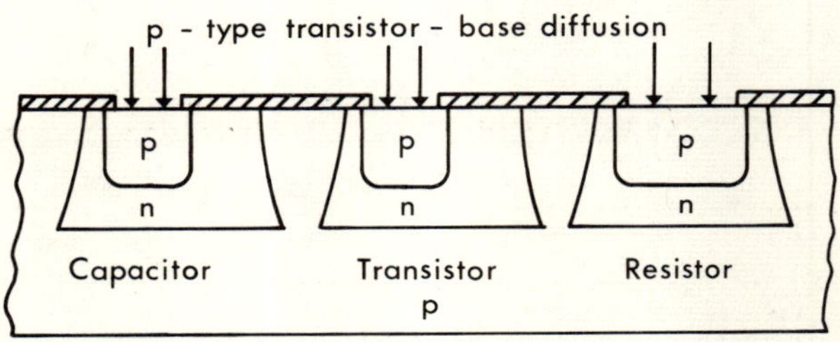

Fig. 7.20 Second diffusion process.

Step 4: A third cycle of oxide removal and diffusion is performed. It is generally referred to as the transistor-emitter diffusion because of the influence of the emitter on the diffusion details. Low resistivity n^+-type regions are produced which form the emitter and the collector contact region of the transistor and the second half of the capacitor (Fig. 7.21). The wafer is re-oxidized.

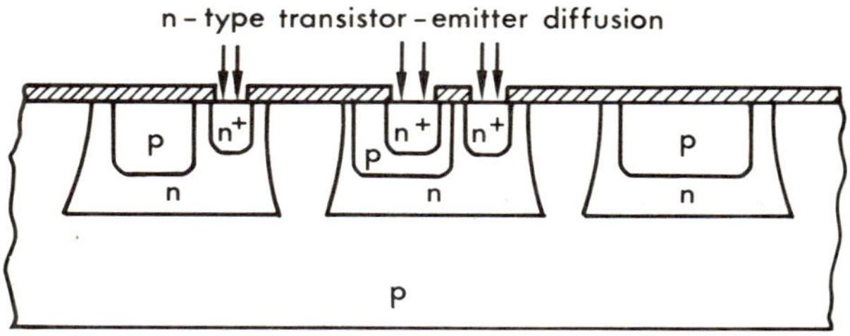

Fig. 7.21 Third diffusion process.

Step 5: Finally, certain regions of the silicon oxide are removed to allow aluminium to form both the electrical contacts with the circuit elements and their interconnection pattern (Fig. 7.22).

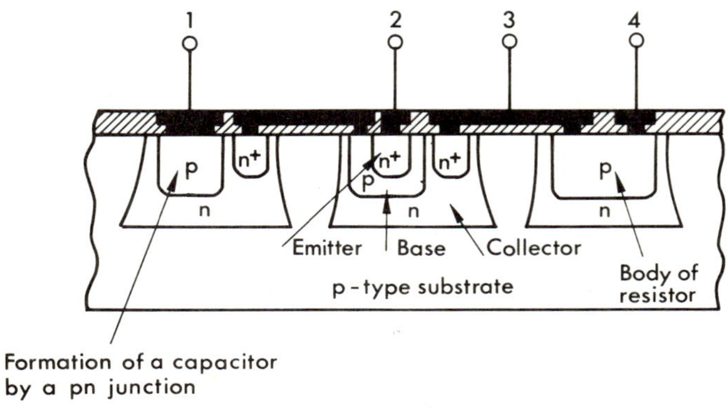

Fig. 7.22 Complete integrated circuit.

7.6 Field Effect Transistors for Integrated Circuits

As mentioned in the previous chapter, the insulated-gate field-effect transistor (or MOST) particularly lends itself to integrated circuit technology. Although the MOST is not suitable for all integrated

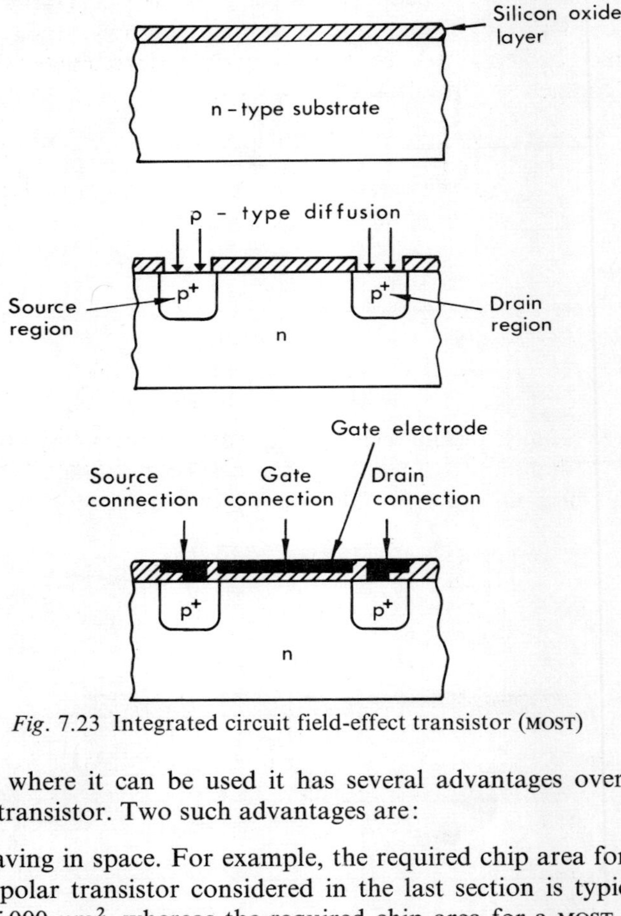

Fig. 7.23 Integrated circuit field-effect transistor (MOST)

circuits, where it can be used it has several advantages over the bipolar transistor. Two such advantages are:

(a) Saving in space. For example, the required chip area for the bipolar transistor considered in the last section is typically 65 000 μm^2, whereas the required chip area for a MOST may be as little as 3200 μm^2.
(b) Saving in cost. For various reasons the fabrication of a MOST is simpler, and this results in a cheaper device.

Fig. 7.23 depicts the various steps involved in the fabrication of an integrated circuit field-effect transistor, (MOST). It may be seen that the fabrication is considerably simpler than the bipolar transistor. The low resistivity (p^+) source and drain regions are formed simultaneously by p-type diffusion into a lightly-doped n-type substrate. After the surface of the chip has been re-oxidized, holes are etched in order for contacts to be made to the source and drain regions. The actual gate electrode is deposited simultaneously with these contacts.

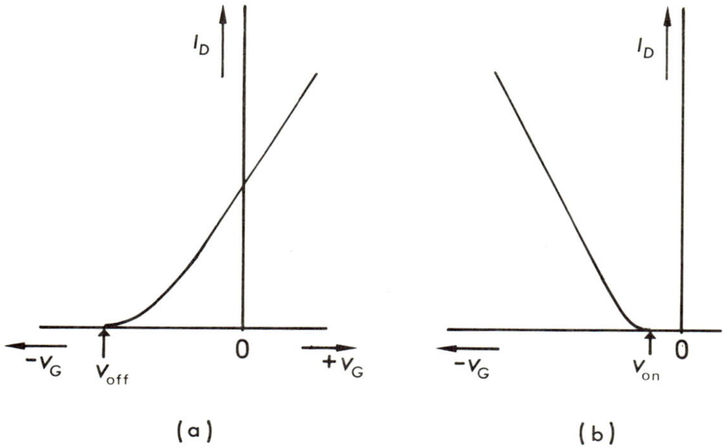

Fig. 7.24 Transfer characteristic of an n- and p-type MOST.
(a) n-channel MOST.
(b) p-channel MOST.

Integrated MOSTs, which are extensively employed for digital work, are generally p-type channel devices because they have enhancement characteristics, i.e. the transistor is 'off' when the gate potential is zero (Fig. 7.24).

Diffused resistors, particularly high resistance ones, take up considerable chip area. For example, a 30 kΩ diffused resistor will occupy approximately three times the chip area of a bipolar transistor. For high value resistances a considerable saving in chip area can be achieved by using a suitably connected MOST. Moreover,

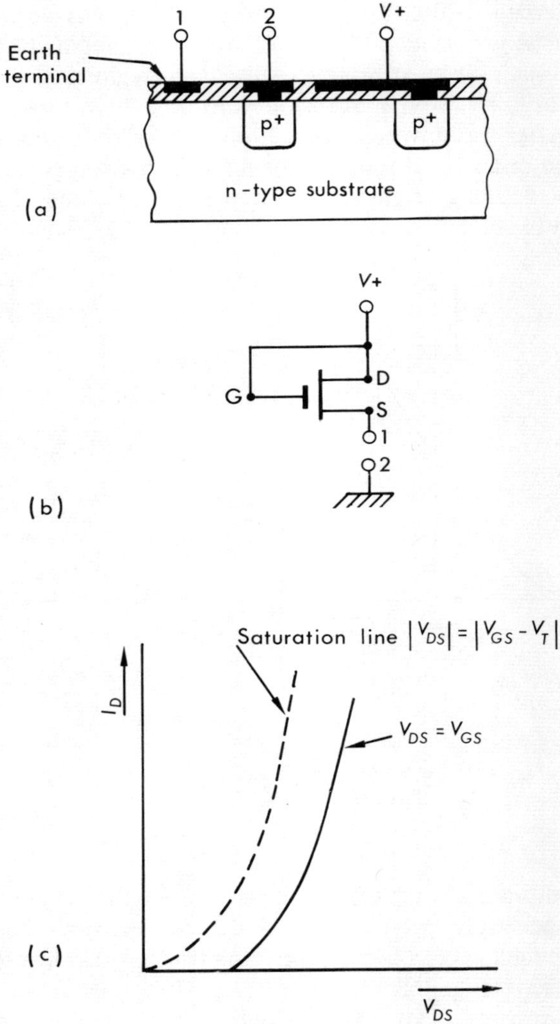

Fig. 7.25 MOST connected as a resistor.
(a) Cross section of structure.
(b) Effective circuit.
(c) I_D/V_{DS} characteristic.

resistances greater than 30 kΩ can be obtained. The resulting saving in chip area using a MOST resistor instead of a diffused resistor for values over 30 kΩ is typically 95 per cent, i.e. the diffused resistor occupies approximately twenty times the chip area of a MOST resistor. Fig. 7.25(a) and (b) show a MOST connected as a resistor.

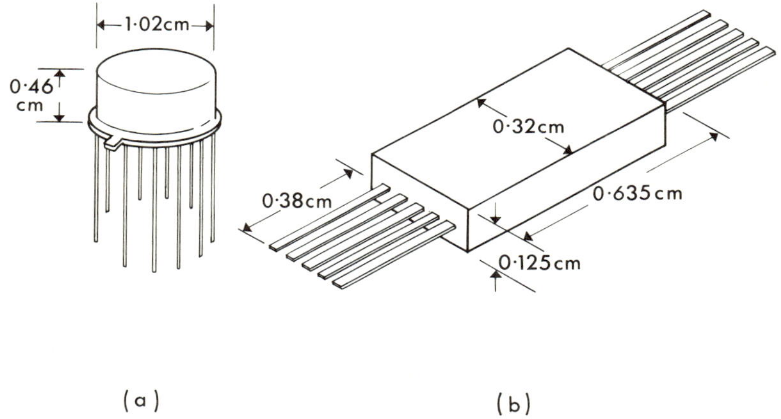

Fig. 7.26 Integrated circuit encapsulation.
(a) TO-99-can.
(b) Flat pack.

The gate is short circuited to the drain to ensure that the transistor always operates in saturation, i.e. saturation occurs for all drain voltages greater than $|V_{GS}-V_T|$. Thus by directly connecting the gate to the drain V_{DS} will always be equal to V_{GS}. This means that V_{DS} will always be greater than $|V_{GS}-V_T|$ ensuring that the transistor will always be operating in saturation. Fig. 7.26(c) shows the effect of this direct connection between the gate and the drain on the relationship between the drain current and drain voltage. The reciprocal of the slope of this slightly non-linear characteristic gives the resistance of the MOST resistor. It can be shown to be approximately equal to $1/g_m$.

7.7 Encapsulation

When all the circuits have been fabricated, the completely processed wafer may contain up to 300 identical chips, each chip having an approximate area of 1·25 mm^2 and a thickness of 0·25 mm. Before the wafer is cut into individual chips, each of the circuits is probe tested. This involves using a testing machine that has a cluster of fine probes which are precisely aligned over and then lowered onto each circuit in turn. The probes make electrical contact with the circuit at certain strategic points and thereby make possible a number of voltage and current measurements. By evaluating these measurements, any faulty circuit may be suitably marked and later rejected when the slice has been cut up into individual chips.

After the chips have been separated and thoroughly tested, they are mounted in a suitable container. Such a container or package must provide the following:

(a) Good reliable interconnection between the minute internal circuit and its external connections.
(b) Compactness and mechanical strength to withstand rough handling.
(c) Suitable heat dissipation, i.e. the heat developed within the chip must be adequately dissipated.
(d) Hermeticity, i.e. protection of the chip against chemical contamination.
(e) As low a cost as possible with regard to the above.

Two widely used forms of encapsulation are shown in Fig. 7.26. They are:

(a) *TO-99-type can.* This is a small cylindrical can approximately 1 cm in diameter with eight or more leads emerging from its base.
(b) *Flat pack.* As its name suggests this is flat rectangular package with ten or more flattened leads emerging from two opposite sides.

Once the chip has been mounted in, say, a TO-99-type can by alloying it to the metal surface of the header, interconnections are made between the appropriate aluminium output points on the chip and the pin connections of the cap. This is performed by attaching very fine gold or aluminium wires by thermocompression bonding. A number of different ways may be employed. One method is for one end of an interconnecting wire (typically 25 μm in diameter) to be brought to bear on the appropriate output contact of the circuit. Then by raising the temperature of the metals towards their annealing values and by applying pressure by means of a wedge-shaped tool, the wire is bonded to the aluminium contact point. A similar process may be used for securing the other end of the wire to the respective output pin of the header.

Encapsulation is complete by welding a metal cap to the header. This seals the circuit against mechanical damage and chemical contamination.

In conclusion, the circuit is thoroughly tested for quality assurance.

REFERENCES

BURGER, R. M. and R. P. DONOVAN (Ed.) *Fundamentals of Silicon Integrated Device Technology, Vol. 1 Oxidation, Diffusion and Epitaxy*, Prentice-Hall, 1967.

Integrated Circuits, Inst. of Elect. Engrs., 1967.

MOTOROLA, INC. *Integrated Circuits*, McGraw-Hill Book Company, 1965.

RUNYAN, W. R. *Silicon Semiconductor Technology*, McGraw-Hill Book Company, 1965.

Glossary of Important Terms

Acceptor impurity. A trivalent impurity which accepts an electron from a neighbouring atom and thus creates a mobile hole.

a.c. resistance. The resistance of an element to a small change of current.

Active device. A device having gain or control.

Anode. The positive terminal of a device.

Atom. The smallest part of an element which can enter into chemical combination.

Avalanche breakdown. Reverse voltage breakdown due to electrons gaining sufficient velocity to dislodge other electrons and thus creating more current carriers.

Base. The region of a transistor situated between the emitter and the collector.

Breakdown voltage. The reverse voltage at which the current begins to increase very rapidly with voltage.

Cathode. The negative terminal of a device.

Channel. The region between source and drain in a FET.

Collector. The end region of a transistor which is normally reverse-biased with respect to the base.

Covalent bonds. Chemical bonds or ties between atoms formed by the interchange of valence electrons between atoms.

Crystal lattice. The regular pattern in which atoms arrange themselves in a crystal.

Current carriers. There are two types of current carriers in semiconductors; electrons and holes. Both are mobile, and under the influence of a voltage they are caused to move in opposite directions due to their negative and positive charge. Their movement constitutes an electric current.

Depletion layer. A carrier-free region in a semiconductor, caused by an electric field.

Diode. A two-electrode device which has an asymmetrical *V/I* characteristic.

Donor impurity. A pentavalent impurity which donates a free electron.

Doping. The addition of impurities to a pure semiconductor.

Drain. A p- or n-type region situated at one end of the channel in a FET.

Electron. A negatively charged particle contained in an atom which orbits around the positively charged nucleus.

Electron–hole pair. When a covalent bond is broken, both an electron and a hole are made available as current carriers.

Element. A substance which cannot chemically be decomposed into simpler substances.

Emitter. The end region of a transistor that is forward biased with respect to the base.

Epitaxy. The growth of material on a semiconductor which assumes the same crystal structure as the starting semiconductor material.

Extrinsic semiconductor. A semiconductor whose electrical characteristics depend upon the presence of impurities.

Forward bias. An external voltage applied to a pn junction which reduces the height of the potential barrier in the depletion layer.

Gate electrode. The control electrode of a FET or SCR.

Hole. A mobile vacancy amongst the valence electrons in a semiconductor. It has a charge equal and opposite to that of an electron.

IGFET. An insulated-gate field effect transistor, i.e. an active semiconductor device in which the conductivity of the channel is controlled by a transverse electric field which is induced capacitively across an insulator.

IGFET, **depletion-type.** An IGFET in which the source, channel and drain have the same type of doping.

IGFET, **enhancement-type.** An IGFET in which the channel is doped with opposite type material to the source and drain.

Impurities. Small traces of pentavalent or trivalent material which are added to pure silicon to change it to n- or p-type silicon.

Integrated circuit. A whole inseparable circuit which is fabricated within a single substrate.

Intrinsic semiconductor. A pure semiconductor containing no impurity atoms.

Ion. If a complete atom gains an electron it is called a negative ion, if it loses an electron it is called a positive ion.

Junction FET. An active semiconductor device in which the conductivity of the channel between source and drain is controlled by the transverse electric field created by a reverse-biased pn junction.

Leakage current. The current which flows across a reverse-biased pn junction.

Majority carriers. The type of carrier which predominates in the current flow.

Microelectronics. A branch of electronics concerned with the design, fabrication and application of extremely small electronic components, circuits and systems.

Minority carriers. The type of carrier which contributes less than half of the current flow.

Mobility. The average drift velocity of charge carriers per unit electric field.

Monolithic. Formed within a single face of material.

MOST. See IGFET.

n-type silicon. Silicon in which the number of free electrons greatly outnumber the holes.

n^+-type silicon. Extra highly doped n-type silicon.

Nucleus. The central part of an atom which carries a positive charge.

Passive device. A device not having gain or control.

Pentavalent atom. An atom with five valence electrons.

Photoresist. An emulsion which changes its chemical solubility when exposed to light.

Pinch-off voltage. The voltage which must be applied to the gate electrode of a depletion-type IGFET to reduce conduction between the source and the drain to zero.

Planar transistor. A transistor which is fabricated by selective etching and diffusion and whose electrodes form essentially parallel planes.

Potential barrier. A small potential which opposes the diffusion of majority carriers across a pn junction due to the charge of the fixed donor and acceptor ions.

p-type silicon. Silicon in which the number of holes greatly outnumber the electrons.

p^+-type silicon. Extra highly doped p-type silicon.

Recombination. When a hole is filled by an electron, recombination has taken place, i.e. the opposite procedure to electron-hole pair generation.

Reverse-bias. An external voltage applied to a pn junction to raise the height of the potential barrier in the depletion layer.

SCR. A silicon controlled rectifier is a four layer pnpn, three-terminal device which can be made to switch rapidly from the non-conducting to the conducting state by a suitable signal applied to the gate electrode.

Semiconductor. A solid or liquid substance whose resistivity is between that of electrical conductors and insulators (10^{-4} to 10^{7} Ωm) and in which the electrical charge carrier density increases with rising temperature over a specific temperature range.

Source. A p- or n-type region situated at one end of the channel in a FET.

Substrate. The foundation or supporting material upon which an integrated circuit is fabricated.

Threshold voltage. The voltage which must be applied to the gate electrode of an enhancement-type IGFET to initiate conduction between the source and the drain.

Thyristor. See SCR.

Transistor. An active semiconductor device having three electrodes.

Transistor, bipolar. A transistor which makes use of both positive and negative charge carriers.

Transistor, unipolar. A transistor whose operation depends on charge carriers of one polarity only.

Triac. A semiconductor device which has a forward and reverse characteristics similar to the forward characteristic of an SCR.

Trivalent atom. An atom with three valence electrons.

Valence electrons. The electrons in the outer shell of an atom which determine the relative ability of one atom to combine with another atom.

Varactor. A two-electrode pn device with a voltage-dependent non-linear capacitance.

Zener diode. A junction diode designed to breakdown at a given voltage.

Index

Acceptor impurity, 23
Amplification factor, 104
Amplification:
  audio frequency, 92
  bias, 88, 118
  load line, 84
Alpha, 54, 177
Avalanche breakdown, 34

Beta, 69
Bias:
  common emitter, 88
  common source, 118
Bootstrap circuit, 129
Breakdown:
  avalanche, 34
  bipolar transistors, 69
  diodes, 34
  IGFET, 155
  junction FET, 101
  Zener, 38
Bridge rectifier, 42
Bypass capacitor, 92

Capacitance:
  depletion, 35
  IGFET, 155
  junction FET, 107
  Miller, 109
  parallel plate, 35
Carrier:
  majority, 23
  minority, 23
Characteristics:
  bipolar transistor, 60–78
  diode, 33
  IGFET, 149, 152
  junction FET, 100, 103
Collector:
  dissipation, 87
  electrode, 50
Coupling capacitor, 92
Covalent bonds, 17
Crystal:
  lattice, 17
  pulling, 200
Current, flow of:
  bipolar transistor, 56
  diode, 28, 30
  IGFET, 141, 142
  junction FET, 99
  leakage, 34, 71
  SCR, 172

Depletion:
  capacitance, 35
  IGFET, 142
Depletion layer:
  bipolar transistors, 51
  diode, 27
  IGFET, 144
  junction FET, 96
Deposition, metal of, 206
Diffusion, 204
Diode:
  equivalent circuit, 37
  integrated circuit, 214
  isolation, 207
  regulator, 46
  resistance, 37
Donor impurity, 21
Doping, 21
Drain electrode, 95

Electrometer, 161
Electron:
  free, 18
  valance, 16
Element isolation, 207
Emitter:
  current, 54
  electrode, 51
  follower, 74
Encapsulation, 230
Enhancement IGFET, 139
Epitaxial growth, 202
Equivalent circuit:
  bipolar transistor, 78
  diode, 36
  IGFET, 151
  junction FET, 93
  SCR, 172
Etching, 204

Fabrication:
  depletion type IGFET, 142
  enhancement type IGFET, 139
  integrated circuit elements, 209
  junction FET, 95
  SCR, 169
  triac, 192
Field effect transistor, 93
Forward:
  bias, 27
  resistance, 34
Free electron, 18

Gain:
  amplifier, 89
  bipolar transistor, 78
Gate:
  characteristic, 180
  electrode, 93, 141, 173
Germanium, 15

Holes, 18
Half-wave rectifier, 40
Hybrid equivalent circuit, 80

Impurities, 21
Ingot, 202
Input capacity:
  IGFET, 155
  junction FET, 109
Input resistance, 55
Inversion, 190
Ions, 22
Isolation of elements:
  oxide, 207
  diode, 207

Junction:
  capacitance, 218
  diodes, 35
  FET, 93

Lattice, 17
Leakage current, 34, 117, 169
Load line, 84

Majority carrier, 23
Mean free path, 112
Metal-oxide semiconductor, 141
Microelectronics, 198
Minority carrier, 23
Miller effect, 109
Mobility, 112
Monolithic integrated circuit, 198
Mutual conductance, 102, 151

npn transistor, 50
n-type semiconductor, 21
nucleus, 15

Off-state SCR, 173
On-state SCR, 173
Operating point, 86
Orbit, 15
Output characteristic:
  bipolar transistor, 61, 68
  IGFET, 147
  junction FET, 99
Oxidation, 203
Oxide isolation, 207

Peak forward voltage, 186

Peak reverse voltage, 187
Peak surge current, 186
Phase shift, 72
Photomasking, 204
Photoresist, 204
Pinch-off effect, 99, 144
pn junction, 26
pnp transistor, 55
Polycrystalline structure, 200
Potential barrier, 26
Proton, 15
Purification of silicon, 200

Quiescent point, 84

Ratings of:
  bipolar transistor, 81
  IGFET, 155
  junction FET, 115
  SCR, 185
RC amplifier, 90
Recombination, 19
Rectification:
  bridge circuit, 42
  centre tapped transformer, 42
  half-wave, 40
Regulator circuit, 46
Reliability of ICs, 199
Reverse bias, 30
Reverse characteristic, 33
Resistance:
  diode, 37
  sheet, 220

Saturation, 99
Seed crystal, 200
Semiconductor:
  crystalline structure, 17
  conduction, 19, 22
  n-type, 21
  p-type, 23
Sheet resistance, 220
Silicon:
  oxide, 203
  controlled rectifier, 168
Smoothing, 44
Source electrode, 95
Source-follower amplifier, 127
Switch:
  IGFET, 162
  SCR, 189

Temperature effect:
  bipolar transistor, 70
  junction FET, 109
  SCR, 181
Thermal compression bonding, 231
Thermal runaway, 87
Thyristor, 168
Transadmittance, 102
Transconductance, 102
Transistor:
  bipolar, 50
  IGFET, 139
  junction FET, 93
Transfer characteristic:
  bipolar transistor, 62, 69
  IGFET, 150
  junction FET, 102

Valence electrons, 16
Varactor, 36
Voltage regulator, 46
Voltage variable resistor, 135

Wafer, 202

Zener breakdown, 38
Zener diode, 38